Ocean Engineering & Oceanography

Volume 7

Series editors

Manhar R. Dhanak, Florida Atlantic University, Boca Raton, USA
Nikolas I. Xiros, New Orleans, USA

More information about this series at http://www.springer.com/series/10524

Arthur Pecher · Jens Peter Kofoed
Editors

Handbook of
Ocean Wave Energy

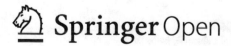 Springer Open

Editors
Arthur Pecher
Wave Energy Research Group, Department
  of Civil Engineering
Aalborg University
Aalborg
Denmark

Jens Peter Kofoed
Wave Energy Research Group, Department
  of Civil Engineering
Aalborg University
Aalborg
Denmark

ISSN 2194-6396   ISSN 2194-640X (electronic)
Ocean Engineering & Oceanography
ISBN 978-3-319-81990-7   ISBN 978-3-319-39889-1   (eBook)
DOI 10.1007/978-3-319-39889-1

The original version of the book was revised:
For detailed information please see erratum.
The erratum to this book is available at
DOI 10.1007/978-3-319-39889-1_11

Preface

This Handbook for Ocean Wave Energy aims at providing a guide into the field of ocean wave energy utilization. The handbook offers a concise yet comprehensive overview of the main aspects and disciplines involved in the development of wave energy converters (WECs). The idea for the book has been shaped by the development, research, and teaching that we have carried out at the Wave Energy Research Group at Aalborg University over the past decades. It is our belief and experience that it would be useful writing and compiling such a handbook in order to enhance the understanding of the sector for a wide variety of potential readers, from investors and developers to students and academics.

At the Wave Energy Research Group, we have a wide range of wave energy related activities ranging from teaching at master and Ph.D. level, undertaking generic research projects and participating in specific research and development projects together with WEC developers and other stakeholders. All these activities have created a solid background in terms of theoretical knowledge, experimental and numerical modeling skills as well as a scientific network, which is why we found that the idea of putting this book together seemed realistic. With this as a starting point, we gathered a group of authors, each an expert within their specific research topic. It was clear from the beginning that the ambition was to make a high-quality publication but still ensuring that it would have a high level of accessibility. Therefore, we wanted the book to be freely available in digital form. To make this happen, we sought and received funding from the Danish EUDP program (project no. 64015-0013), for which we are extreme thankful.

The ten chapters of the handbook present a broad range of relevant rules of thumb and topics, such as the technical and economic development of a WEC, wave energy resource, wave energy economics, WEC hydrodynamics, power take-off systems, mooring systems as well as the experimental and numerical simulation of WECs. It covers the topic of wave energy conversion from different perspectives, providing the readers, who are experts in one particular topic, with a clear overview of the key aspects in other relevant topics in which they might be less specialized.

We would especially like to thank our co-authors, who have contributed enthusiastically to the content and without whom we would never have been able to realize this handbook. We would also like to thank our colleagues at the Department of Civil Engineering for supporting us, especially Kim Nielsen who patiently helped us getting all the small final details in place as well as reading through all the chapters for final corrections and comments, and Vivi Søndergaard who gave the final touch to the English language.

Last but not least, we would like to thank our wives, Marie Isolde Müller and Kirsten Aalstrup Kofoed, for their endless patience and support.

We have enjoyed working with you all and we are very grateful for each of your contribution.

Aalborg, Denmark   Arthur Pecher
2016   Jens Peter Kofoed

Contents

Abbreviations

AEP	Annual Energy Production
BoP	Balance of Plant
CapEx	Capital Expenditure
CF	Cash Flow or Capacity Factor
CFD	Computational Fluid Dynamics
CHP	Combined Heat and Power
CoE	Cost of Energy
CRL	Commercial Readiness Level
CWR	Capture Width Ratio
Dec	Decommissioning
DevEx	Development Expenditure
DoF	Degree of Freedom
DPP	Discounted Payback Period
EMEC	European Marine Energy Centre
FCF	Free Cash Flow
FV	Future Value
IRF	Impulse Response Function
IRR	Internal Rate of Return
IW	Irregular Waves
LCoE	Levelized Cost of Energy
MAEP	Mean Annual Energy Production
NPV	Net Present Value
OpEx	Operational Expenditure
OWC	Oscillating Water Columns
PI	Profitability Index
PLC	Programmable Logic Controller
PP	Payback Period
PTO	Power Take-Off
PV	Present Value
RADR	Risk Adjusted Discount Rate

RANSE	Reynolds-Averaged Navier-Stokes Equation
R&D	Research and Development
RAO	Response Amplitude Operator
RW	Regular Waves
SS	Sea State
TPL	Technology Performance Level
TRL	Technology Readiness Level
VoF	Volume of Fluid
WAB	Wave Activated Body
WACC	Weighted Average Cost of Capital
WEC	Wave Energy Converter

Symbols

a	Amplitude of incident wave (m)
A	Cross-sectional area (m^2)
$A(\omega)$	Frequency-dependent added mass (kg)
A_c	Effective cross-sectional area of a pair of cylinders (m^2)
A_e	Electric loading (A/m)
AEP	Annual energy production (MWh/year)
a_i	Amplitudes at each frequency (m)
A_{wp}	Water plane area (m^2)
A_∞	Limiting added mass coefficient at infinite frequency (kg)
B	Air gap magnetic flux density (Tesla) or center of buoyancy (m)
B_{e11}	Equivalent damping coefficient (surge) (–)
C	Shape coefficient (–)
c	Wave celerity (m/s)
c_a	Speed of sound in atmospheric conditions (m/s)
C_d	Wave drift force coefficient (–)
C_D	Drag coefficient (–)
C_g	Group velocity (m/s)
CI	Confidence interval (–)
C_m	Added mass coefficient (–)
$Contrib$	Contribution to the available wave power (–)
D	Damping coefficient (kg/s)
D_b	Float draft below the water surface (m)
D_{es}	Damping constant for the end stop mechanism (kg/s)
D_h	Float height above surface (m)
D_t	Turbine rotor diameter (m)
E	Electromotive force (V)
F	Fetch length (m)
f	Frequency (Hz)
F	Force (N)
F_3	Restoring force (N)

F_b	Buoyancy force (N)
F_c	Current force (N)
F_e	Excitation force (N)
f_{exc}	Excitation impulse response function (–)
F_{exc}	Excitation force (N)
F_f	Friction force (N)
F_{hs}	Hydrostatic force (N)
F_m	The Mooring force (N)
F_{pto}	PTO force (N)
F_r	Radiation force (N)
$f_w(f)$	Wave force ratio (–)
G	Center of gravity (m)
g	Gravitational acceleration (m/s^2)
G	Hydrostatic matrix (N/m)
γ	Peak enhancement factor (–)
G_m	Constant (–)
GZ	Righting arm (m)
h	Water depth (m)
H	Wave height (m), heaviside step function (–) or horizontal force (N)
$H_{1/3}$	Significant wave height (m)
H_{max}	Max wave height within a given duration of a sea state (m)
Hp	Horizontal pretension (N)
H_s	Significant wave height (m)
$h(t-\tau)$	Impulse-response function (–)
H_{m0}	Significant Wave Height estimate from wave spectrum (m)
I	Current density in the conductor (A)
I_o	Incident momentum (–)
i,j	Mode of motion (–). Translations: 1: Surge, 2: Sway, 3: Heave. Rotations: 4: Roll, 5: Pitch, 6: Yaw
J	Wave power flux or wave power level (equal to P_{wave}) (kW/m)
K	Roughness height (mm)
k	Spring coefficient or Stiffness (N/m)
k	Wave number (m^{-1})
k/D	Relative roughness (–)
KC	Keulegan–Carpenter number (–)
K_{es}	Spring constant for the end stop mechanism (N/m)
K_t	Constant that depends only on turbine geometry (–)
l	Length (m)
L_l	Leakage inductance (H)
L_m	Main inductance (H)
$L_{p,0}$	Wave length based on peak wave period and deep water (m)
L_s	Synchronous inductance (H)
m	Body mass (kg)
m	Mass (kg)
M	Mass matrix (kg)

$MAEP$	Mean annual energy production (MWh/year)
m_0	Variance of the wave spectra or 'zeroth' moment of the wave spectra (m^2)
m_n	Spectral moment of the nth order (n = 0, 1, 2, ...) ($m^2\ s^{-n}$)
m_r	Added mass (kg)
$m^{\cdot}$	Mass flow rate of air through the turbine (kg/s)
m_n	Spectrum moments (–)
N	Number of coil turns (–)
N	Number of harmonic wave components (–)
N or $\acute{\omega}$	Rotational speed (radians per unit time) (rad/s)
N_c	Number of pairs of cylinders (–)
N_L	Length scaling factor (–)
p	Differential pressure in the pneumatic chamber (Pa)
p_a	Atmospheric pressure (Pa)
P_{abs}	Primary absorbed power from the waves (kW)
$P_{available}$	Available power (kW)
P_{el}	Generated electrical power (kW)
P_{mech}	Available mechanical power (kW)
$Prob$	Probability of occurrence (–)
P_t	Turbine power output (kW)
P_u	Useful power (kW)
P_{wave}	Wave power flux or wave power level (equal to J) (kW/m)
Q	Volume flow rate of liquid displaced by the piston (m^3/s)
q	Volume flow rate of air (m^3/s)
q_0	Mass per unit unstretched length (kg/m)
Q_m	Flow rate (m^3/s)
r	Amplitude of reflected wave (m)
R	Damping (kg/s)
Re	Reynolds number (–)
R_g	Resistance inside the generator (Ω)
R_l	Resistance (Ω)
S	Stiffness (N/m), spectral density function (m^2/Hz) or scaling ratio (S)
s	Wave steepness or sample standard deviation (–)
S_f	Spectral density at frequency component f (m^2/Hz)
S_{mbs}	Minimum breaking strength (N)
$S_{p,0}$	Wave steepness for the peak wave period and deep water (–)
t	Time or amplitude of transmitted wave (s) or (m)
T	Wave period or wave record with duration (s)
T_0	Resonance period (s)
T_{01}	Spectral wave period based on 0th and 1st moment (s)
T_{02}	Spectral wave period based on 0th and 2nd moment (spectral estimate of T_z) (s)
T_B	Breaking load (N)
T_e	Wave energy period (s)
T_p	Peak wave period (s)

T_{QS}	Quasi static tension (N)
T_z	Mean zero down crossing wave period (s)
u	Horizontal water velocity (m/s) or usage factor (–)
U	Velocity (m/s)
U_{10}	Wind speed at a height of 10 m (m/s)
$U_{10min,10m}$	Mean wind speed over 10 min at 10 m height (m/s)
U_c	Current speed (m/s)
U_f	Full scale velocity (m/s)
U_m	Model scale velocity (m/s)
u_{max}	Maximum water velocity (m/s)
u_r	Relative speed (m/s)
V	Volume (m^3)
V_s	Available stroke volume (m^3)
w	Distance between the poles (m)
x	Horizontal position of the body (m)
X_c	Quasi static line extension (m)
$\dot{x}$	Velocity of the body (m/s)
$\ddot{x}$	Acceleration of the body (m/s^2)
z	Vertical displacement (m)
$\dot{z}$	Vertical velocity (m/s)
λ	Wave length (m)
Δf	Frequency interval (Hz)
Δp_c	Pressure difference between the accumulators (–)
Φ	Flow coefficient (–)
Π	Power coefficient (–)
Ψ	Pressure coefficient (–)
α_i	Phases of each frequency (Hz)
β	Wave direction (degree)
ϕ	Permanent magnet induced flux per pole or Constant for fixed entropy (B) or (–)
γ	Specific heat ratio for the gas (–) or peak enhancement factor (–)
μ_0	Magnetic permeability (H m^{-1})
ν	Specific volume of gas (m^3) or kinematic viscosity (m^2/s)
ρ	Density (kg/m^3)
ρ_{cu}	Resistivity of the conductor material (Ω)
ω	Angular frequency (rad/s)
ξ	Acceleration vector (m/s^2)
∇	Submerged volume (m^3)
$\zeta(z)$	Vertical displacement of the water particles (m)
$\xi(z)$	Horizontal displacement of the water particles (m)
η	Free surface elevation (m) or non-dimensional performance (also called CWR or efficiency) (–)
$\dot{\eta}$	Velocity of water surface (m/s)
$\ddot{\eta}$	Acceleration (m/s^2)
η_3	Body displacement (m)

η_i	Position in mode (m)
η_{lim}	Excursion limit for which end stop mechanism starts acting (m)
$\eta_{overall}$	Overall non-dimensional performance (efficiency) (–)
η_{PTO}	PTO efficiency (–)
η_{ss}	Non-dimensional performance (efficiency) in individual sea state (–)
η_{w2w}	Wave-to-wire efficiency (–)
ε_0	Spectral bandwidth (–)

Chapter 1
Introduction

Arthur Pecher and Jens Peter Kofoed

1.1 Introduction

The widespread usage of affordable electricity converted from ocean waves would be a fabulous achievement. Besides that the wave energy converting (WEC) technology would be particularly interesting, it also would have several significant benefits to society, such as:

- It is another sustainable and endless energy source, which could significantly contribute to the renewable energy mix. In general, increasing the amount and diversity of the renewable energy mix is very beneficial as it increases the availability and reduces the need for fossil fuels.
- Electricity from wave energy will make countries more self-sufficient in energy and thereby less dependent on energy import from other countries (note: oil is often imported from politically unstable countries).
- It will contribute to the creation of a new sector containing, innovation and employment.
- Electricity from ocean wave can be produced offshore, which thereby does not require land nor has a significant visual impact.

As the world energy needs will keep on increasing while the fossil fuel reserves are depleting, wave energy will become of significant importance. The demand for it will start when its price of electricity will be right and will then only increase with time.

A. Pecher (✉) · J.P. Kofoed (✉)
Department of Civil Engineering, Aalborg University,
Thomas Manns Vej 23, 9220 Aalborg Ø, Denmark
e-mail: apecher@gmail.com

J.P. Kofoed
e-mail: jpk@civil.aau.dk

© The Author(s) 2017
A. Pecher and J.P. Kofoed (eds.), *Handbook of Ocean Wave Energy*,
Ocean Engineering & Oceanography 7, DOI 10.1007/978-3-319-39889-1_1

1

1.2   The Successful Product Innovation

In general, there are three key elements to a successful product innovation. It has to be technically feasible, economically viable and desirable/useable by an end-user. In other words, it requires a new functional technology that has a positive business case and that is of use for society. These key elements do not necessarily require being developed at the same time since a developer needs to start somewhere. However, they need to be present in some kind of harmony before an innovation can successfully be launched on the market (Fig. 1.1).

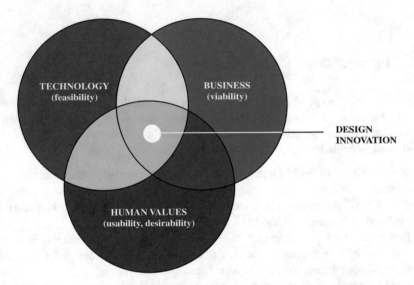

Fig. 1.1 The three key elements of successful product design innovation. Inspired by [25–27]

There is a great demand for renewable energy and a need to diversify the renewable energy mix. This can easily be seen on the significant annual increase in global investment in renewable energy, such as wind and solar. Wave energy has even been additionally stimulated in some countries as they recognise its benefits and great potential. The technology push came mainly in the form of public grants and capital investment in technology development, while the market pull through public market incentives, such as revenue support (the feed-in tariffs) [1, 2]. This indicates that the usability and desirability (or human value) are currently very positive.

An impressive amount of wave energy technologies have been developed over the last 25 years. To give an indication hereof, the list of current wave energy developers at EMEC counts 256 developers [3]. The working principles of most of these technologies can be grouped into a handful of main categories. This just indicates how great the effort has been from the developers (see more in Chap. 2).

The last missing factor for production innovation success is the business potential or economic viability of the wave energy technologies within the frames of the market (with or without incentives). The business case is made based on cost (CapEx and OpEx) and power production calculations (read more in Chaps. 4 and 5). To be able to demonstrate a positive business case, a significant amount of proof (for the calculation) and thereby experience with the WEC is expected to be gathered before. Although some investors can be convinced on the way in the great business potential of a WEC, it will probably still require a decent track record of an offshore full-scale WEC before it will convince a larger market. This is particularly difficult to realise with WECs since the development cost is particularly high (e.g. compared to wind energy) and the development process long. This is especially due to the harsh offshore environment, which requires special equipment and vessels and which is not easily accessible. So, the development process requires a careful balance between technology optimisation and physical progress. The best advice is, therefore, to keep on investigating the economic potential along the development progress as there is no reason to progress if it is absent.

1.3 Sketching WECs and Their Environment

WECs are machines that are able to exploit the power from ocean waves and to convert it into a useable form of energy, such as electricity.

Ocean waves are theoretically relatively well understood and extensively described in literature. However, in practice, it is very difficult to accurately describe, reproduce and predict the exact environmental conditions at a certain offshore location. This is due to its complexity and the large amount of environmental parameters that can have a significant influence on it (read more about this in Chap. 3).

In Fig. 1.2, the different metocean parameters affecting the marine environment are sketchily presented, together with the primary sub-systems of a (floating) WEC.

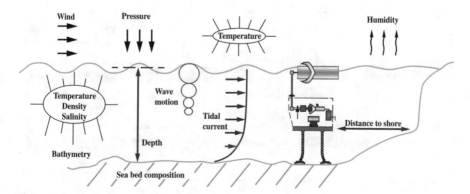

Fig. 1.2 Metocean parameters applicable to marine energy converts, and their primary sub-systems. Adapted from [4]

Most WECs, even the ones with different working principles (see Chap. 2) are very similar from a generic point of view. Most of them consist of the same primary sub-systems, which is due to their common environment and goal (Fig. 1.3).

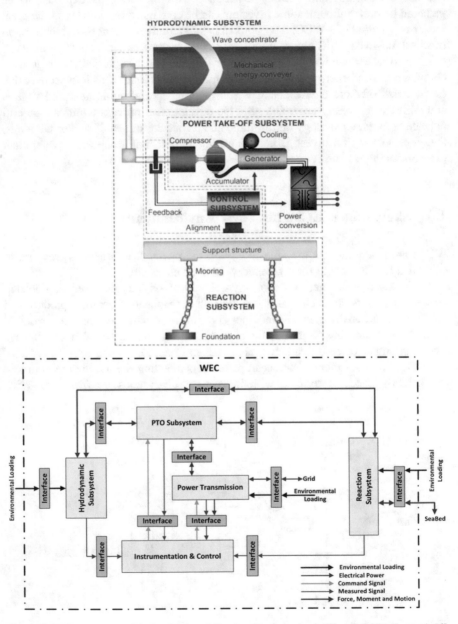

Fig. 1.3 WEC system design breakdown following Equimar (*top*) [5] and DNV (*bottom*) [6]. Courtesy of Equimar and DNV GL

The main sub-systems that are present in (all) WECs have also been introduced widely in literature [4–6] and consist of:

- *The hydrodynamic subsystem* is the primary wave absorption system that exploits the wave power (see Chap. 6). It can be of different types depending on the technology, e.g. oscillating body, oscillating water column and overtopping principle, and it is connected to both the reaction and PTO subsystems against which it will actively transfer forces and motions.
- *The power take-off subsystem* converts the captured wave energy (by the hydrodynamic subsystem) into electricity (see Chap. 8). The PTO systems can be based on different principles, of which some of the most common are hydraulic PTO, direct drive mechanical PTO, linear generators, air turbine and low head water turbine.
- *The reaction subsystem* maintains the WEC into position relative to the seabed (e.g. mooring system) and provides a reaction point for the PTO and/or support for the hydrodynamic subsystem(s) (e.g. fixed reference or support structure) (see Chap. 7).
- *The control (and instrumentation) subsystem* is the intelligent part of the system as it takes care of the control of the WEC and its measurements. It mainly consists of the processors for the automation and electromechanical processes, the sensors and their data acquisition, the communication and data transfer, and the human interface.

These different sub-systems and their interconnections can be presented in different manners, of which two are presented in Fig. 1.3.

1.4 Rules of Thumb for Wave Energy

The following list of "rules of thumb"—covering the essential features, the economics, the design, the PTO systems and the environment of WECS—contains a series of condensed and critical indications which are considered valuable in the assessment of a WEC technology and project. All of them will be addressed in more details in the following chapters.

1.4.1 The Essential Features of a WEC

The following features are the **essential aspects in which a WEC should excel** in order to show long-term economic potential [7]:

- *Survivability*: The WEC requires a reliable mooring system and preferably a passive safety system that can effectively reduce extreme loads. With passive meaning that the safety mechanism can be activated (automatically) without requiring external interaction, such as electricity or other.

- *Reliability and maintainability*: Easy access and inspection of the most essential parts of the WEC. In addition, it would be very beneficial if most (or all) maintenance could be done on the WEC itself at location, without having to bring it back to a harbour.
- *Overall power performance*: The WEC must consist of an efficient wave energy absorbing technology and PTO. It has to produce a sufficiently smooth electrical power and have a high capacity factor. Otherwise, too much energy will be lost over the whole wave-to-wire power conversion chain.
- *Scalability*: At full scale, a WEC needs to be a multi-MW device in order to be economically viable. In order to be able to continue significantly improving its LCoE, it needs to be scalable, meaning that it should be capable of further enlarging its dimensions (like offshore wind turbines do). Many WECs unfortunately reach their optimal dimensions at too low dimensions, making it not possible for them to become multi-MW WECs (>5 MW). This does not include the multiplication of WECs as this will not have a significant influence on the average infrastructural and technology costs and thereby will not significantly improve the LCoE of the WEC or project.
- *Environmental benefit*: WECs are expected to be sustainable energy systems and are thereby expected to have a great environmental benefit and a minimal environmental footprint.

1.4.2 Economic Rules of Thumb

1. For an offshore wind turbine in a 1000 MW farm at 30 m of water depth, an indication of related costs are (more details can be found in Chap. 5 and [8, 9]) as follows:

   - The CapEx per installed MW is approx. 4 million euros.
   - The OpEx/MWh is approximately 30 Euro.
   - The LCoE is approximately 120 Euro/MWh.
   - The general development, infrastructure and commissioning costs, referred to as the base CapEx, of a 3.6 MW offshore wind turbine in a project are in the range of 7.2 million Euros. This includes the development and consent, the installation and commissioning and a part of the balance of plant category, but excludes the tower, the foundations and the technology itself. This cost corresponds to about 45 % of the CapEx [10].
   - The resulting "base" CapEx cost for a 3.6 MW WEC is expected to be slightly less, approx. 6 million Euro, as especially the installation cost should be significantly lower. For smaller WECs, it is expected to be approximately 2 million Euro for a 750 kW WEC.

2. A fast, but reasonably accurate (± 50 %), **estimation of the annual energy production** (*AEP*) of a WEC can be obtained by multiplying the mean wave power level (P_{wave}) with the width of the absorber, the overall wave-to-wire efficiency (η_{w2w}, which is the weighted average over all the wave conditions), the availability and the yearly production hours:

$$AEP = P_{wave} \times width_{absorber} \times \eta_{w2w} \times availability \times hours_{annual}$$

As an example, for a well-functioning optimized point absorber in a good wave environment, this could give (these indicative values used here are set more in context on other following rules of thumb):

$$AEP = 40\,\text{kW/m} \times 15\,\text{m} \times 20\,\% \times 95\,\% \times 8766 = 999\,\text{MWh/year}$$

This corresponds to an average power production of 114 kW, which gives an installed capacity of 750 kW with a capacity factor of 15 %.
The economic value of this is 150 kEuro/year, assuming a feed-in-tariff of 150 Euro/MWh.
If we assume a WEC that is 10 $\times$ larger, we can expect (following the same calculation) that the power production and thereby the revenue will be 10 $\times$ larger as well. Furthermore, it can be expected that the capacity factor will be significantly higher, e.g. 30 % or approx. 3.6 MW, as the capacity factor of WECs improves with the amount of wave absorbing bodies that are connected to the same system (see Table 1.3). This is because the different units will significantly smoothen the overall absorbed power as the different absorbers will have a time offset between the moment in which the different absorbers interact with the same wave, and thereby the max-to-mean power ratio is significantly lower of a common PTO system.

3. Combining the base CapEx cost (does not include the technology itself, nor the OpEx) and the revenue from these two different sizes of WEC, **it will take the small WEC about 13 years to repay its base CapEx cost**, while it will only take **about four years for the large WEC**. This indicates clearly that WECs need to be large to be (-come) economically viable, meaning in the multi-MW scale (≥ 1 MW). The assumption that multiple small WECs can be equally as good as one large WEC does not make economic sense as it is too much challenged by the costs of the base CapEx, meaning the project development, infrastructural and commissioning costs.

4. Besides sharing the base costs more efficiently, large WECs have as well multiple other advantages such as:

- Sharing basic equipment over different wave absorbing bodies, such as mooring systems, weather stations, communication systems, electricity cables and others.
- Sharing parts of the power take-off (PTO) system, which (usually) results into higher capacity factors and smoother electrical power output.

- The whole system can be commissioned at once, thereby sharing installation and servicing works and equipment, e.g. it only requires one vessel for handling one system.
- Larger structures are more easily accessed as they are more stable, which enables easier inspection of the system and some maintenance could be done on board, without the need of retracting the system to a safe/controlled area.

5. There are various technical assessment ratios for a WEC:

- The *wave-to-wire efficiency* (η_{w2w}) is the overall efficiency of the system delivering the absorbed energy from the waves to the grid. This value is also based on many underlying specifications, such as the wave conditions, the availability of the system and the maximum power rating, and so needs to be taken very carefully.
- The *capture width ratio (CWR)* describes the effectiveness of the converter to absorb the energy in the waves. This value is based on many underlying specifications, such as the wave conditions and the size of the wave activated body, and needs thereby to be handled very carefully.
- The *WEC weight/installed kW* ratio is also often used to indicate how much material is used relative to the power rating of the WEC. This can be a bit misleading as it does not particularly show the type of material (e.g. steel or concrete). It should at least be divided between active structural (load carrying) material and ballast material, as their difference in cost can be as great as a factor 100.
- The *capacity factor* (also called capacity factor) is the ratio between the average produced power and the installed power on the WEC. It describes the utility rate of the PTO system and is very interesting as it gives an idea of what the WEC delivers (average produced power) and what it costs (driven by installed power). However, this value is also wave condition dependant (location).

Note that the overall efficiency of a WEC η_{w2w} includes the efficiencies of each power conversion step, between wave and grid, together with the limitations of the system, such as the saturation of the generator. The complete power conversion train is, thereby, composed of at least: hydrodynamic conversion (wave to absorber described by CWR), PTO (absorber to generator), generator and electronics, substation and voltage increase, and grid connection. The availability of the system is not calculated in the overall efficiency as it is dependent on other aspects such as the maintenance possibilities of the system, but is included in the capacity factor.

6. **A very important long-term economic aspect of a WEC is its capability of being scalable in size**, even after it reaches commercial maturity. This can be compared with wind turbines, which keep on being increased in size in order to reduce their LCoE. Different wave-absorbing bodies have different optimal dimensions (see Table 1.2), e.g. the hydrodynamic optimal full-scale diameter of a point absorber will (normally) be between 15–20 m depending on the wave conditions. Large structures with multiple wave absorbing bodies could possibly increase their amount instead of enlarging them.

1.4.3 WEC Design Rules of Thumb

1. The ability for a body to absorb the energy in the waves depends upon its hydrodynamic design (for more details refer to Chap. 6). In general, it can be said that [11]:

"A good wave absorber must be a good wave-maker."

This means that when a body moves in the water, it will create a wave depending on its shape and motion = radiated wave, e.g. a point absorber will make a circular wave equal in all directions when oscillating vertically. The better that this radiated wave corresponds to the incoming ocean wave, the more efficient this body is in absorbing an incoming ocean wave (Fig. 1.4).

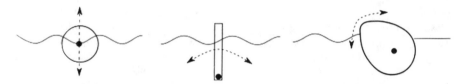

Fig. 1.4 Illustration of the radiated wave by the motion in one direction by three wave-absorbing bodies, from *left to right*: heaving point absorber, pitching flap and pitching Salter's duck

The theoretical limit in wave energy absorption by a body that creates an (anti-) symmetrical radiated wave (e.g. heaving point absorber and pitching flap) is of 50 %. However, for a non-symmetric body (such as a Salter's duck), it may have the ability to absorb almost 100 % incoming wave energy [12].

2. Although there is no clear convergence in technologies yet, there are different main WEC categories. For some of these main categories, **an indicative capture width ratio** on the absorbed power from the waves can be given, based on a collection of published results [13] (Table 1.1).

Table 1.1 Overview of the mean capture width ratio for some of the main WEC types

WEC type	Capture width ratio (%)
Floating overtopping device	17
Oscillating water column	29
Point absorber	16
Pitching flap (bottom fixed)	37

These numbers present a rough indication of the ability of these WEC types to absorb wave energy. This energy still needs to be converted into electricity afterwards. Note that these values need to be taken with care as they can be based on different specifications and assumptions. Some of the most influential parameters are the wave conditions and the relative size (scaling ratio) of the WEC to the waves.

3. **The optimal dimension of the wave absorbing body** and structure of a WEC
   is usually most strongly linked to the wave period (from all the wave parameters), besides other potentially interfering economic parameters. The peak wave
   period with the highest annual wave energy contribution (corresponding to the
   wave energy x probability of occurrence) should be taken into account for this.
   Table 1.2 gives a rough indication of these dimensions for a full-scale WEC in
   an average suitable offshore location [13–17].

Table 1.2 Indication of hydrodynamic optimal full-scale dimension of certain WEC technologies
for average northern European wave conditions

WEC type	Relevant dimension (m)	
Point absorber	Diameter	12–20
OWC	Length[a]	12–20
OWSC	Thickness[a]	The thicker the better
Floating structures e.g. overtopping WEC	Length	Longer than a wavelength

[a]The width of these wave-activated bodies can be chosen independently, but they still have a
strong influence on their hydrodynamic response as it influences the inertia, added mass, drag
coefficient and possibly other characteristics of the wave-activated body. However, they tend to be
in the range of 12–20 m

These values can indicate the scaling possibilities of a full-scale WEC type and,
thereby, indicate the limit in power absorption by a WEC as well.

4. **The power fluctuations of a single WEC** decrease significantly with its
   amount of wave energy absorbers. The absorbed power from waves fluctuates
   due to the nature of the waves (time scale of a few seconds), but also due to the
   fact that waves travel in groups (time scale of a few minutes). These fluctuations
   are not desirable as they increase the need for oversizing mechanical and
   electrical equipment and are one of the main barriers to achieve a reliable and
   cost-efficient technology [18]. Typical max-to-mean ratios in absorbed power
   are (over 1000 waves period, without physical limitations) [19–22] as follows
   (Table 1.3).

Table 1.3 Indication of the max-to-mean ratio on the absorbed power by a WEC with different
configurations

WEC type	Max-to-mean ratio
Single wave-activated body with one-way PTO	15–30
Single wave activated body with two-way PTO	10–12
OWC with two way PTO	10–15
10 side-by-side located wave-activated bodies (in the wave direction) with two-way PTO	3–7

5. As with wind turbines, several sub-system failures should be expected annually, of which an extensive survey on the **failure rates of several subsystems** of wind turbines is given in (Fig. 1.5). In general, due to serious improvements in the last 5–10 years, although wind turbines endure a high number of malfunctions corresponding they normally only lead to short standstill periods due to the rapid interaction of service teams. They achieve a technical availability of about 98 %, corresponding to a downtime of about 1 week a year [23]. This should clearly indicate that it is of high importance that all the vital/critical

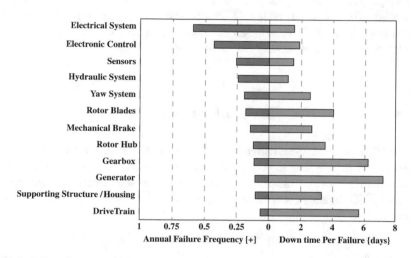

Fig. 1.5 Failure frequency and downtime of components. Adapted from [23]

components of WEC should be at least easy to inspect as several malfunctions will occur every year. Even better would be that the components of the WEC are easy to maintain and to interchange, without the necessity of requiring divers or of bringing the WEC back to a protected environment (e.g. harbour). These are both very expensive, require good weather windows, are unpractical and are time-consuming. Fully submerged WECs are, thereby, really difficult to operate as their maintainability is very difficult (remembering that the WEC is located in an area with serious wave conditions).

6. For WEC technologies having a main floating reference structure, it is desirable that the projected length of such a WEC is approximately the same or more than a wavelength for optimal power production. In the opposite case where the wavelength is much longer than the structure, the structure will start moving with the wave.

7. **Mooring of floating structures** can be problematic and is in general expensive. Some basic rules of thumb are as follows:

   • Although WECs are typically more efficient in steep waves, they result in larger surge offsets, relative to their rest position.

- Surge motions of a moored floating structure are especially large under the event of breaking waves, which also result in significantly higher wave loads on the structure.
- The durability of the mooring system is even further challenged under short-term repetitive wave events such as wave groups (which is very common).

A golden rule is to moor a floating WEC outside of the area where wave breaking occurs due to water depth interferences.

8. **Exceptionally high (peak) loads occur with sudden stops of bodies in motion**. This can occur within the structure or sub-systems, e.g. due to physical end-stops in the PTO system (e.g. in linear generators or hydraulic pistons) or snap shocks when stretching out mooring lines.

1.4.4 Power Take-Off Rules of Thumb

1. The **PTO of a wave-activated body is the most efficient when its motion is restricted to only one degree of freedom**. Otherwise, the wave-activated body will always chose to move in the direction of the least resistance and thereby avoid PTO interaction. Furthermore, limiting its motions to one degree of freedom;

   - Reduces the complexity of the PTO system and the possible amount of load cases.
   - Optimises its efficiency and facilitates its control as the exact motion of the wave activated body is known.

2. PTO systems for WECs are normally required to convert a slow oscillating movement combined with high forces (induced by the nature of the waves) to a fast rotation in one direction (required by an electrical motor). Thereby, there is a wide range of different types of PTO systems, which all present advantages and inconveniences in term of efficiency, control, complexity and cost (see Chap. 8). **Indicative values of efficiencies for these different types of PTOs** (from absorbed wave energy to generator) are [24] (Table 1.4).

   Note that other aspects of the PTO system can be as well of high importance, such as the ability to;

Table 1.4 Overview of the indicative efficiency for different PTO systems (see more on Chap. 8)

PTO system	Efficiency (%)
Hydraulic	65
Water	85
Air	55
Mechanical	90
Direct drive	95

- Temporarily store/smooth energy.
- Handle short-term power overload.
- Handle sudden system faults and possible control losses.

3. Advanced control strategies of the wave absorbing body through the PTO system can typically greatly enhance the overall power production. However, this will also entail significantly higher loads and wear on the structure and components of the system.
4. The **PTO is also much more efficient working against a fixed reference**. This fixed reference can be the seabed or a large structure that does not move under the wave absorbing action of the system. Otherwise, a lot of energy will potentially be transferred into motions of other linked bodies.

1.4.5 Environmental Rules of Thumb

1. **The power performance of a WEC is much better in steep waves** as these results in more frequent and/or larger motions of the wave energy absorber. In long (swell) waves, the motions of the water surface are less frequent and slower, which lead to slower and smaller motions of the wave-activated body.
2. **Important aspects of a good location for WECs**:

   - Good average wave energy content, e.g. >15 kW/m, as this is the source of energy.
   - Good average wave steepness, e.g. >1.5 %, as the performance of WECs is significantly higher in steep waves.
   - Low max-to-mean ratio in terms of significant wave heights, as you build (≈pay for) the WEC design to endure a 100-year wave while it produces energy (≈earnings) relative to the average wave condition.
   - Low monthly wave energy content variation, as it facilitates stable power production and improves the capacity factor when the wave climate is consistent over the whole year. However, this makes installation and maintenance more difficult as weather windows are less frequent and shorter.
   - Proximity to the coast, infrastructure and end-user as it significantly reduces CapEx and OpEx costs related to the project.
   - Reasonable water depth (e.g. 30–60 m), which can seriously affect the mooring and cabling cost.

It can hardly be expected to find a place where all of these criteria are met perfectly. However, it is the best balance in between them, resulting in the best overall LCoE, which should dictate the value of a project at a certain location. Some of the better locations are the following:

- South and West coasts below the tropic of Capricorn (e.g. Australia, New Zealand, South Africa and Chile): high average wave power and low seasonal variability and low 100-year wave to mean wave ratio.
- East coasts below the tropic of Capricorn (e.g. Australia, New Zealand, South Africa, Argentina, Uruguay and South Brazil): medium average wave power, with low seasonal variability and low 100-year wave to mean wave ratio.
- West coast of United States: medium average wave power, with low seasonal variability and low 100-year wave to mean wave ratio.
- North Atlantic (Europe and East coast US): high average wave power and steep waves, but high seasonal variability and high 100-year wave to mean wave ratio.

References

1. SI OCEAN: Wave and Tidal Energy Market Deployment Strategy for Europe, p. 47 (2014)
2. COWI: The Potential of Market Pull Instruments for Promoting Innovation in Environmental Characteristics, pp. 1–110 (2009)
3. EMEC: EMEC Webpage. http://www.emec.org.uk/marine-energy/wave-developers/. Accessed 17 Jan 2016
4. Myers, L., Bahaj, A., Retzler, C., Sørensen, H., Gardner, F., Bittencourt, C., Flinn, J., Equimar Delivrable D5.5: Guidance on pre-deployment and operational actions associated with marine energy arrays. EquiMar Protocols—Equitable Testing and Evaluation of Marine Energy Extraction Devices in terms of Performance, Cost and Environmental Impact (2010)
5. Hamedni, B., Ferreira, C.B., Cocho, M.: Generic WEC System Breakdown (2014)
6. SI Ocean Project: Ocean Energy: State of the Art Technology Assessment (2012)
7. Pecher, A., Kofoed, J.P., Larsen, T.: The extensive R & D behind the Weptos WEC. RENEW, Lisbon, Portugal (2014)
8. Deloitte: Establishing the investment case wind power (2014)
9. Christensen, B.: Den nødvendige indsats i offshore vindindustrien - Siemens Wind Power, aarsmoedet 2014 i offshoreenergy.dk (2014)
10. BVG Associates for the Renewables Advisory Board.: Value breakdown for the offshore wind sector (2010)
11. Falnes, J., Budal, K.: Wave power conversion by point absorbers. Nor. Marit. Res. **6**(4), 2–11 (1978)
12. Falnes, J.: Principles for capture of energy from ocean waves. Phase control and optimum oscillation. Department of Physics, NTNU, N-7034 Trondheim, Norway (1997)
13. Babarit, A.: A database of capture width ratio of wave energy converters renew. Energy, **80**, 610–628 (2015)
14. Pecher, A.: Performance Evaluation of Wave Energy Converters. Aalborg University (2012)
15. Gomes, R.P.F., Lopes, M.F.P., Henriques, J.C.C., Gato, L.M.C., Falcão, A.F.O.: The dynamics and power extraction of bottom-hinged plate wave energy converters in regular and irregular waves. Ocean Eng. **96**, 86–99 (2015)
16. Henry, A., Doherty, K., Cameron, L., Whittaker, T., Doherty, R.: Advances in the Design of the Oyster Wave Energy Converter, pp. 1–10 (2011)
17. Faltinsen, O.M.: Sea loads on ships and offshore structures (1990)
18. Bjørnstad, E.: Control of wave energy converter with constrained electric power take off, NTNU (2011)

19. Sjolte, J., Sandvik, C., Tedeschi, E., Molinas, M.: Exploring the potential for increased production from the wave energy converter lifesaver by reactive control. Energies 6(8), 3706–3733 (2013)
20. Sidenmark, M., Josefsson, A., Berghuvud, A., Broman, G.: The Ocean Harvester—Modelling, Simulation and Experimental Validation, Simulation, pp. 421–425 (2009)
21. Pecher, A., Kofoed, J.P.: Experimental study on the updated PTO system of the WEPTOS wave energy Converter. Aalborg University DCE Contract Report 138 (2013).
22. Sjolte, J., Bjerke, I., Hjetland, E., Tjensvoll, G.: All-Electric Wave Energy Power Take Off Generator Optimized by High Overspeed, pp. 2–5 (2011)
23. Hahn, B., Durstewitz, M., Rohrig, K.: Reliability of Wind Turbines: Experiences of 15 years with 1,500 WTs. In: Wind Energy Proceedings of Euromech Colloquium, pp. 329–332 (2007)
24. Chozas, J.F., Kofoed, J.P., Helstrup, N.E.: The COE Calculation Tool for Wave Energy Converters (Version 1.6, April 2014) (2014)
25. Brown, T.: Design innovation. d.School Stanford (2005)
26. Scholz, C.: Think, Make, Do—Principles of Design Thinking (2014). https://www.linkedin.com/pulse/20140723095828-11028866-think-make-do-principles-of-design-thinking. Accessed 18 Jan 2016
27. The Foundation for Enterprise Development (FED): Business Plan Creation: A Unified Process for Businesses with a Government and Commercial Focus, DARPA Small Bus. Programs Off (2013). http://dtsn.darpa.mil/sbpt/help/help.html. Accessed 18 Jan 2016

Chapter 2
The Wave Energy Sector

Jens Peter Kofoed

2.1 Introduction

When entering the field of wave energy utilization it is relevant to ask—why is it important to start utilizing this resource? The reasons for this are shared with other renewable energy sources, such as hydro, wind, solar, biomass and other ocean energy forms such as tidal, currents, thermal and salinity driven systems. The key issues that the use of renewable energy sources can help to overcome includes environmental problems, depletion of the fossil fuels, security of supply and job creation.

The environmental problems relates to both local effects such as pollution but also the production of CO_2, which is related to energy production using fossil fuels, with the now well established negative effects on climate change as a consequence [1].

The depletion of fossil fuels was already highlighted in publications in the 1950s [2] and it is well established that the fossil fuels are finite and that the time horizon before they are depleted are counted in 10'ths, maybe 100'ths, of years. Thus, it is also obvious that the current level of energy consumption, which is by far majority based on fossil fuels, cannot continue unless alternative sources are developed. And here the renewable energy sources are the most obvious answers, as these resources will be available as long as the sun is shining.

But even still while there currently are reasonable amounts of fossil fuels available, the uneven distribution of the resource around the globe is giving rise to conflicts. It can only be expected that this tendency will be worsened as the fossil

J.P. Kofoed (✉)
Department of Civil Engineering, Wave Energy Research Group, Aalborg University,
Thomas Manns Vej 23, 9220 Aalborg Ø, Denmark
e-mail: jpk@civil.aau.dk

© The Author(s) 2017
A. Pecher and J.P. Kofoed (eds.), *Handbook of Ocean Wave Energy*,
Ocean Engineering & Oceanography 7, DOI 10.1007/978-3-319-39889-1_2

resources are getting more and more depleted. Thus, for most nations it is of great interest to decrease their dependency on fuel supply from other countries to maintain their sovereignty and political stability. As an answer to that renewable energy sources are very diverse and to a much larger extent scattered and well distributed around the globe, when looking at the renewable energy resource as a whole. Locally, large variations are present in which kind of renewable energies it is relevant to utilize. This also means that there is a need to develop a broad portfolio of renewable energy technologies to have sufficient 'tools in the toolbox' to fit the local needs.

In the current market, energy from the less mature technologies utilizing renewable energy sources are generally not cost competitive, but relies on political support. However, it can be expected that this situation will turn in the near future due to both the expected (and experienced) increase in cost of fossil fuels and the reduction of cost of the technologies utilizing renewable energy sources, due to further R&D and economics of scale.

In Denmark, as an example, it is a political goal to make the country independent of fossil fuels by year 2050. In September 2010 the Danish Commission on Climate Change Policy presented its suggestions as to how Denmark in the future can phase out fossil fuels [3]. The Commission's work had to reflect the ambition of the European Union that developed countries should collectively reduce their emissions of greenhouse gases by 60–80 % by 2050.

The task of the Commission was to present proposals for new proactive instruments for an energy and climate change policy with global and market-based perspectives that contribute to cost-effective attainment of the long-term vision. The Commission also had to assess new fields of technology and the potential for the market-based development of these technologies with the aim of implementing the long-term vision and assess the extent to which effective implementation requires internationally coordinated cooperation.

The analysis carried out by the Commission substantiates that a conversion of the energy system to be 100 % independent of the fossil fuels in 2050 is a realistic goal. Costs to society of such a conversion will only be modest.

To realize this goal a number of initiatives and technologies needs to be deployed. The key elements are; more efficient use of energy and the energy has to come from renewable sources. To get an impression of what is suggested refer to Fig. 2.1. A major part of the supply of electricity is expected to come from wind (60–80 % compared to 20 % today), but as illustrated in Fig. 2.1, also wave energy is foreseen as a technology contributing to the future energy mix. In other countries the wave energy is expected to a take a more central role—this is tightly linked to the available resource. As illustrated later in this chapter, the wave power level on the European Atlantic west coasts is up to 5 times greater than in the Danish part of the North Sea.

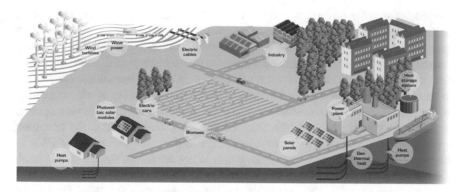

Fig. 2.1 An energy system without fossil fuels [22]

Compared to fossil fuel based energy production technologies most renewable energy investments are spent on materials and workmanship to build and maintain the facilities, rather than on fuel, which for renewable energy sources are for free. Renewable energy investments are to a much larger extent spent within the nation and often in the same local area (as a large share of the cost is going to operation and maintenance of the facilities) where the produced energy is consumed. This means the investment to a large intent stays in the neighbourhood where it create jobs and fuel the local economies, rather than going to regions far away. On the other hand, there are also opportunities for export for the nations who are successful in developing commercially viable renewable energy technologies. This has been experienced extensively in Denmark in where wind turbines are now one of the most important export articles.

When talking about utilization of wave energy for electricity production the current state is that the technologies are not yet mature. A number of full scale demonstration projects exists (examples hereof are given later in this chapter), but these are generally still in the R&D phase and the cost of the produced energy from these installations are multiple times greater than the target (market) level. However, efforts to reduce costs through optimization of structures, operation, control etc. as well as economics of scale are expected to be able to bring the cost of energy down to a level which at least is comparable with other more mature renewable energy technologies (such as offshore wind). It is in this context the current publication should be seen and is contributing to the advance of the field of wave energy utilization.

2.2 Potential of Wave Energy

When considering wave energy as a source for electricity production it relevant and interesting to look at the estimates of how large the potential for utilization is. Ocean waves including swells (waves generated by distant weather systems) are derived from solar energy, through wind, which when blowing over the ocean surface generates the waves. The waves travel over great distances with very little energy loss, as long as the waves are in deep water conditions. The types of ocean surface waves considered when taking about wave energy utilization are further discussed in Chap. 3. Here it is just noted that when addressing utilization of ocean waves what is meant is the wind generated waves (and possibly swells, depending on the specific device characteristics). Thus, the scope is limited to looking at ocean surface waves with periods in the range of 0.5–30 s. Besides tides, the remaining wave types hold in practice no potential for utilization. The utilization of tides for energy production is termed tidal energy and is not addressed further in this book.

When considering sea states (characterized by statistical wave parameters covering e.g. periods of ∼1000 waves) these are steadier than the wind field which generates the waves. The wave energy flux (power level) exhibits significant variation in time and space. It can range from a few W/m up to MW/m in extreme (stormy) conditions. The wave power level also exhibits a significant seasonal variation (1:5 in Danish waters), as well as year-to-year variation (±50 % in Danish waters) [4].

Early estimations of the global available wave power indicates a total potential of 2.7 (−70) TW [5]. [6] present a more detailed an updated study of the world wide wave energy potential, illustrated in Figs. 2.2 and 2.3 broken down into regions of the world.

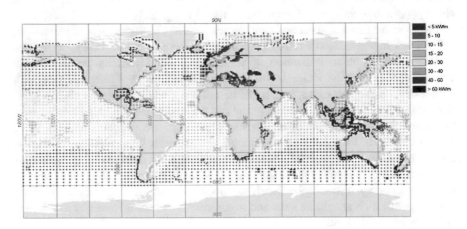

Fig. 2.2 Annual global gross theoretical wave power for all WorldWaves grid points worldwide [6]

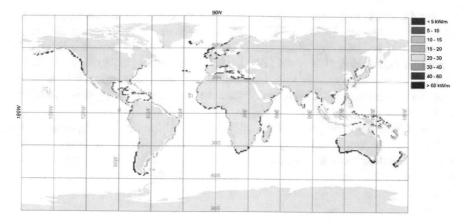

Fig. 2.3 Annual net theoretical coastal power worldwide (excluding contributions where $P \leq 5$ kW/m and potentially ice covered areas) [6]

The global gross theoretical resource is estimated at about 3.7 TW, 3.5 TW is the resource computed excluding areas with a benign wave climate (areas with less than 5 kW/m) and the net resource (where also areas with potential ice cover is excluded) is about 3 TW; the total reduction from gross to net resource is then about 20 %. In Europe there is a decrease of 25 % from gross to net resource, mostly a result of ice coverage, the gross and net values being 381 and 286 GW, respectively. To put these numbers into context note that the total world consumption in 2008 was 142.300 TWh [7] corresponding to an average power of 16.2 TW. In terms of electricity consumption the corresponding numbers are 20261 TWh and 2.3 TW [8]. Thus, the total wave energy resource exceeds by far the global consumption of electricity.

In many regions of the world more local studies of the wave energy resource have been performed, see e.g. [9]. Here, not only the gross theoretical resource is estimated for the US, but also the total recoverable resource (under specific assumptions) is estimated to be 44 % of the gross theoretical resource. So, if it is assumed that the numbers for the US can be applied world-wide it is reasonable to expect that the total recoverable global wave energy resource is approx. 2/3's of the global electricity consumption.

For Europe it is suggested in [10] that a total of 100 GW install capacity of ocean energy (note—this includes also a contribution from tidal energy), generating 260 TWh/y, by 2050 is a realistic target. For comparison it can be noted that in 2005 83 TWh was produced by 40 GW of installed wind turbine capacity in Europe, and by 2030 these numbers are expected to reach 965 TWh and 300 GW [11]. In other words, wave energy has a significant potential for Europe, but will most likely remain minor compared to the wind industry. However, as the renewable energy resources cover a larger and larger share of the electricity consumption, the timing and predictability of the power production becomes increasingly important, and in this respect will a combination of wind and wave (in combination with the other renewable energy sources) be far more beneficial compared to wind alone.

For Denmark, a mapping of the wave energy potential in the Danish part of the North Sea is available from [4]. In here, the gross theoretical resource for this area is estimated to be 3.4 GW, which can be compared to the annual electricity consumption in Denmark [12] of 4.4 GW, i.e. 85 % hereof. In [13] a rough estimation of how this resource could be utilized is given, showing a production of 30 % of the electricity consumption is technically feasible (or 35 % of the gross theoretical resource).

So, to sum up—the potential of wave energy utilization for supplying a significant part of world electricity needs is there. Next question is then regarding which technologies can be used for this purpose? A more detailed description of the wave energy resource is given in the dedicated Chap. 3 entitled: Wave energy resource.

2.3 Wave Energy Converters

2.3.1 History

The development of wave energy converters (WEC's) goes far back in time—the first attempts are recorded to have taken place in the 1800s, see Fig. 2.4 and [14]. Actually, the first patent for a wave energy converter dates back the year 1799. In

Fig. 2.4 Postcard from a "wave motor" experiment off the coast of Santa Cruz from 1898 [23]

modern time it was not until the energy crisis in the beginning of the 1970s that the field had renewed interest, greatly boosted by an article by Prof. Stephen Salter in the scientific journal Nature in 1974 [15]. However, in spite of very significant research efforts, not the least in the UK, activities were reduced again up through the 1980s and the beginning of the 1990s. By the end of the past millennium activities were picking up in speed again, and this now in a number of countries around the world, but with most efforts seen in the coastal European countries. Over the past decade UK has again put enormous efforts into development of marine renewable energies, including wave energy, and must today be seen as the world leader in the field.

2.3.2   Categorization of WEC's

The development of WEC's is characterized by the fact that there is a large number of different ideas and concepts for how to utilize the wave energy resource. The different concepts can be categorized in a number of different ways.

Often a basic categorization using the terms terminator, attenuator and point absorber is used [42]. *Terminators* are devices with large horizontal extensions parallel to the direction of wave propagation, while *attenuators* have large horizontal extensions orthogonal to the direction of wave propagation. In contrast *point absorbers* with extensions small compared to the predominant wavelength of the prevailing waves.

WEC's can also be categorized by their location—*onshore, near shore* and *offshore*. Onshore, or shore-mounted, devices are by nature terminators, and rigidly connected to land. Typical examples hereof are oscillating waver columns and overtopping devices, see further explanation below. Near shore devices are situated at water depths where the available waves are influenced by the water depth, and devices deployed in this region will often be bottom mounted. And thus, at last, devices placed offshore will generally be floating and have access to the waves unaltered by the presence of the seabed.

Classification of WEC's is also seen by their main working principles [44]. The European Marine Energy Center at the Orkney Islands is using 8 main types, plus one ('other'—in acceptance of the fact that some WEC's cannot be put into the existing boxes).

However, in the following, the approach to categorization used by IEA—Ocean Energy Systems [16] will be used. This approach is illustrated in Fig. 2.5.

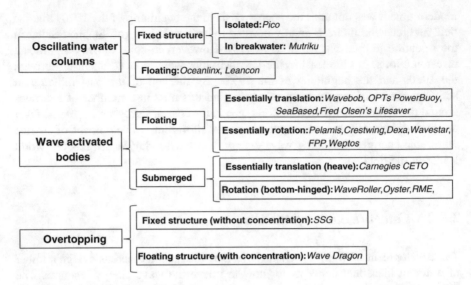

Fig. 2.5 Categorization of wave energy technologies following IEA – Ocean Energy Systems [16]. Technologies mentioned in the various categories are the ones illustrated in the following figures

Here, all WEC's consisting of oscillating bodies are put into one category. This category is termed Wave Activated Bodies (WAB's).

In an attempt to detail the categorization of WEC's a level further, guidelines have been provided by the EU-FP7 funded EquiMar project on how to categorize WEC's by subsystems [17]. The WEC is in this case broken into the following subsystems, which can then be individually categorized:

- Primary energy extraction
- Power take-off/control system
- Reaction system

The concept for detailed categorization and breakdown of a WEC is developed further by DNV-GL in [18], which provides a generic system breakdown useful as base of a generic risk ranking and failure mode analysis.

2.3.3 Examples of Various WEC Types

In the following a wide range of examples of WECs, however only a fraction of the technologies that are currently being developed around the world, are presented, here categorized according to the categories defined by IEA—Ocean Energy Systems (following Fig. 2.5).

2.3.3.1 Oscillating Water Column

There is a number of shore-based (fixed) oscillating water columns (OWC) WECs that has been operating, on Islay in Scotland (operated by *WaveGen*), the Pico plant on the Azores in Portugal (Fig. 2.7), at the port Mutriku breakwater in Spain (Fig. 2.6), Sagata port Japan and OceanLinx Australia (Fig. 2.7). The unidirectional rotation of the air turbine (the Wells type) is a simple way to rectify the bidirectional flow and thereby convert the oscillating power from waves, due to the fact that the need for check-valves can be omitted and the structure thus constructed with less moving parts. Voith Hydro WaveGen Limited has been developing this type of turbines.

Fig. 2.6 Mutriku Oscillating Water Column breakwater, equipped with 16 WaveGen Wells turbines, total capacity of 300 kW [24]

Fig. 2.7 The Pico OWC, schematics (*top*, *left*) and in action (*top*, *right*) [25], OceanLynx, schematics (*lower*, *left*), and in real life (*lower*, *right*) [26]

The *LeanCon* WEC is floating and also based on the concept of oscillating water columns (Fig. 2.8). It is a large structure covering more than one wave length and it consists of a large number of OWC chambers. This entails that the resulting vertical force on the WEC is limited. The downward forces from the negative pressure on parts of the WEC prevent it from floating up on the top of the waves and due to this the device can have a low weight (constructed from high strength fiber reinforced material). Before the air flow reaches the power take off system (PTO) the air flow is rectified by the non-return valves. Thus, LeanCon uses uni-directional air turbine, while most other OWC use Wells turbines.

Fig. 2.8 Leancon 1:40 scale model in wave basin (*top*, *left*), 1:10 prototype for testing in Nissum Bredning, Denmark, under construction (*top*, *right*) and after deployment (*middle* and *bottom*). Courtesy of LeanCon

2.3.3.2 Wave Activated Bodies

The category of wave activated bodies (WABs) encompasses a very large field of WEC concepts. In this section a number of examples are given to give an impression of the plurality, but it cannot be considered complete as the number of concepts in this category can be counted in hundreds.

The *Pelamis* WEC is a floating device, made up of five tube sections linked by universal joints which allow flexing in two directions (Fig. 2.9). The WEC floats semi-submerged on the surface of the water and inherently faces into the direction of the waves, kept in place by a mooring system. As waves pass down the length of the machine and the sections bend in the water, the movement is converted into electricity via hydraulic power take-off systems housed inside each joint of the machine tubes, and power is transmitted to shore using standard subsea cables and equipment [19].

Fig. 2.9 E.ON P2 Pelamis operating in Orkney July 2011 [27]

Like Pelamis, the *Crestwing* is a moored device utilizing the relative motion between wave activated bodies (Fig. 2.10). While Pelamis is harvesting the energy from 2 degrees of freedom (DOF) in a total of 4 joints, Crestwing is just using a single DOF for power production. The hinged rafts of the Crestwing are closed box structures. And the PTO of the Crestwing is a mechanical system using a ratchet mechanism and a fly wheel for converting the oscillatory motion between the rafts into a rotating motion on an axle, which can be fed into a gear and generator system. Other concepts have been tested using relative rotation between floating

Fig. 2.10 Crestwing, at a scale of 1:5, tested near Frederikshavn during autumn 2011 [28]

Fig. 2.11 Picture of Dexawave at DanWEC [29]

bodies includes *Dexa* (Fig. 2.11), *Martifer, MacCabe Wave Pump* and *Cockerell's Raft*.

Another group of floating WABs includes translating (often heaving) bodies. This includes devices *Ocean Power Technologies (OPT)* (Fig. 2.12), which is one among a number of technologies utilizing a point absorber. The OPT PowerBuoy is using a reference plate as point of reference for the PTO. OPT has used different solutions for PTO, including oil hydraulics. OPT is working on a range of deployment projects, and have conducted sea trials using both a 40 kW and 150 kW version of their technology. Other devices using similar approaches include *Wavebob* (using a submerged volume rather than a damping plate for reference) (Fig. 2.13) and *SeaBased* (using a fixed reference point at the seabed, where also the PTO, a linear generator, is placed (Fig. 2.14).

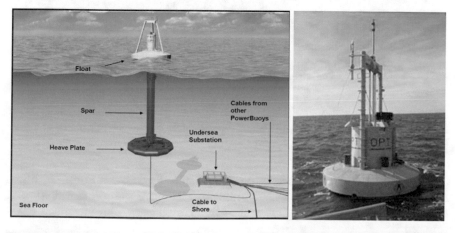

Fig. 2.12 OPTs PowerBuoy PB40. A slack moored pointer absorber with heave plate [30]

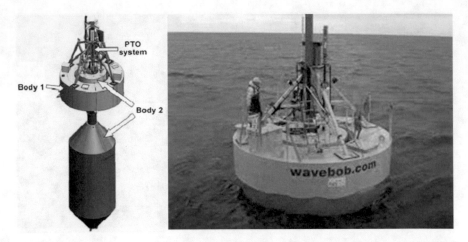

Fig. 2.13 Wavebob. A slack moored pointer absorber with submerged reference volume [31]

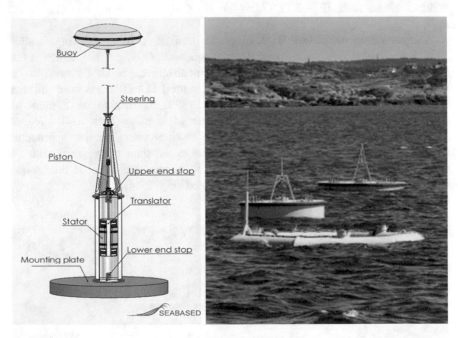

Fig. 2.14 SeaBased. A point absorber with a directly driven linear generator placed at the sea bottom [32]

Other types of point absorbers also exists, such as Fred Olsens Lifesaver, which not only utilizes the heave (translation) but also the pitch and roll (rotation), as it consists of a torus connect to the seabed through winches with integrated PTOs (Fig. 2.15).

Fig. 2.15 Fred Olsen's Lifesaver buoy, illustration (*left*) and deployed at Falmouth test site (FabTest) (*right*). Courtesy of Fred Olsen

As an example of a submerged WAB Carnegies *CETO* buoy can be mentioned. In the CETO device the buoy itself is completely submerged and kept in place by a tether fixed at the seabed and with a hydraulic pump based PTO in line (Fig. 2.16).

Fig. 2.16 Carnegies CETO. A submerged tether moored point absorber [33]

Besides the above, another group of fixed WABs, specifically submerged flaps hinged at the seabed, can be mentioned. This type includes *Oyster*, developed by Aquamarine, which was announced in 2001 by Professor Trevor Whittaker's team at Queens University in Belfast (Fig. 2.17). The flap is moved back and forth by the waves, and power is taken out through hydraulic pumps mounted between the flap and the structure pinned to the seabed. The latest generation Oyster 800 has an installed capacity of 800 kW. It has a width 26 m and height of 12 m was installed in a water depth of 13 m approx. 500 m from the coast of Orkney at EMEC.

Fig. 2.17 Installation of the Oyster 800 submerged flap WEC at the European Marine Energy Centre in Orkney, Scotland [34]

Other relevant WECs utilizing same operating principles includes *Waveroller* (Fig. 2.18), *Resolute Marine Energy* (Fig. 2.19) and *Langlee* (Fig. 2.20). However, the latter is not fixed to the seabed, but a structure with two flaps attached to a floating reference frame.

Fig. 2.18 Waveroller WEC prototype, before submerged to seabed [35]

Fig. 2.19 Resolute Marine Energy (RME) WEC prototype [36]

Fig. 2.20 Artist impression of the Langlee WEC. A floating submerged flap WEC [37]

In addition to the above mentioned WECs in the WAB category, also a number of devices exist where multiple bodies are combined into one larger structure. An example here is the *Wavestar* device, which consists of two rows of round floats—point absorbers—attached to a bridge structure, fixed to the sea bed by the use of steel piles, which are cast into concrete foundations (Fig. 2.21). All moving parts are therefore above normal seawater level. The device is installed with the structural bridge supporting the floats directed towards the dominant wave direction. When the wave passes, the floats move up and down driven by the passing waves, thereby pumping hydraulic fluid into a common hydraulic manifold system which produces a flow of high pressure oil into a hydraulic motor that directly drives an electric generator. A prototype with a total of two floaters (diameter of 5 m) has been undergoing sea trials at DanWEC, Hanstholm, Denmark.

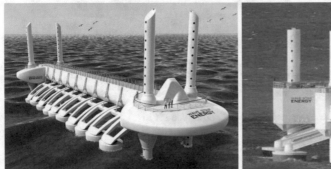

Fig. 2.21 Wavestar prototype with two floaters, at DanWEC, Hanstholm, Denmark (*left*) and concept for full scale deployment, artists impression (*right*). Courtesy of Wavestar [38]

Another multi-body device is the Floating Power Plant (Fig. 2.22). This device is a moored structure utilizing multiple WABs aligned parallel to the wave crests. Thus, the operating principle resembles to some extent the Wavestar, except the reference structure here is floating and not bottom mounted. Furthermore, the floating structure is used as a floating foundation for wind turbines. Floating Power Plant has carried out sea trials at a benign site with a reduced scaled prototype, and is currently preparing its first full scale prototype deployment.

Fig. 2.22 Floating Power Plant, prototype deployed at Vindeby off-shore wind turbine farm, off the coast of Lolland in Denmark (*top*) and illustration of latest design (*bottom*). Courtesy of Floating Power Plant

The *Weptos* WEC is another floating and slack-moored structure, composed of two symmetrical frames ("legs") that support a multitude (20) of identical rotors (Fig. 2.23). The shape of these rotors is based on the shape of Salter's duck WEC (invented and intensively developed since 1974 [15]). All rotors on one leg are connected to the same frame are driving a common axle. Each axle is connected to an independent PTO. The torque, resulting from the pivoting motion of the rotors around the axle, is transmitted through one-way bearings on the up- and down-stroke motion of the rotor. The angle between the two main legs is adaptable. This allows the device to adapt its configuration relative to the wave conditions, increasing its width relative to the incoming wave front in operating wave conditions and reducing its interaction with excessive wave power in extreme wave conditions.

Fig. 2.23 A fully functional WEPTOS model undergoing testing at CCOB, IH Cantabria, Spain (Sept. 2011) (*top*) and artist impression of the full scale Weptos WEC (*bottom*). Courtesy of Weptos [39]

2.3.3.3 Overtopping Devices

The *Wave Dragon* is a slack moored WEC utilizing the overtopping principle (Fig. 2.24). The structure consists of a floating platform with an integrated reservoir and a ramp. The waves overtops the ramp and enters the reservoir, were the water is temporarily stored before it is led back to the sea via hydro turbines generating power to the grid, and thereby utilizing the obtained head in the reservoir. Furthermore, the platform is equipped with two reflectors focusing the incoming waves towards the ramp, which thereby enhance the power production capability.

Other overtopping based approaches do also exist, including the *SSG*, which is a fixed structure acting as a combination of a WEC and a breakwater (Fig. 2.25). In order to still being able to harvest the wave power with good efficiency, while not having the option of adjusting the ramp height through the floating level, SSG consists of multiple reservoirs with different heights. However, simpler approaches with just a single reservoir integrated into (existing) breakwaters are also being explored.

2.3.4 The Development of WECs

As seen above a large variety of WECs exists, and more are still appearing. EquiMar (an EU FP7 funded research project [20]), along with others, has promoted the use of a staged development approach to the development of WEC's, and

Fig. 2.24 Wave Dragon 1:4.5 scale grid connected prototype tested in Nissum Bredning, Denmark

Fig. 2.25 Conceptual drawings of the SSG [40]

thus, the stage of development can also be used for characterization of the WEC's. EquiMar uses 5 stages to describe the development of a WEC from idea to commercial product. These 5 stages are illustrated in Fig. 2.26.

Each stage should provide specific valuable information to inventor and investors, before going to the next step, and hereby avoid spending too many resources before having a reliable estimate on the concepts potential.

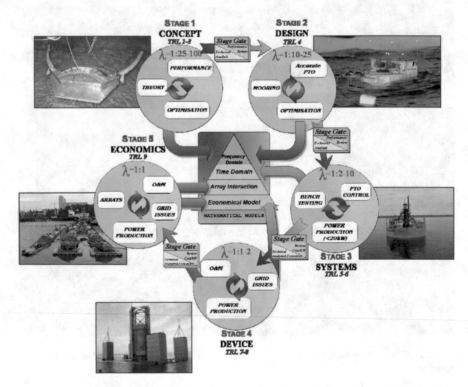

Fig. 2.26 The 5 development stages used for description of the development of a WEC from idea to commercial product used by the EquiMar project. Courtesy of EquiMar [20]

This topic will extensively be addressed in the corresponding Chap. 4 entitled: Techno-economic development of WECs.

As seen from the above examples of WECs and the staged development approach, an important element of the development is the initial real sea testing of the WEC prototype, which paramount prior to commercial introduction to the market. This has called establishment of test sites in real sea, which is the topic for the next section.

2.4   Test Sites

A number of test sites for testing and demonstration of WEC prototypes at real sea have been established throughout Europe over the past couple of decades. One of the first, and most developed, is the European Marine Energy Center (EMEC), established in 2003 at the Orkney Islands, which is providing open-sea testing facilities, as well consultancy and research services.

In Fig. 2.27 this and many other test sites in Europe are pointed out.

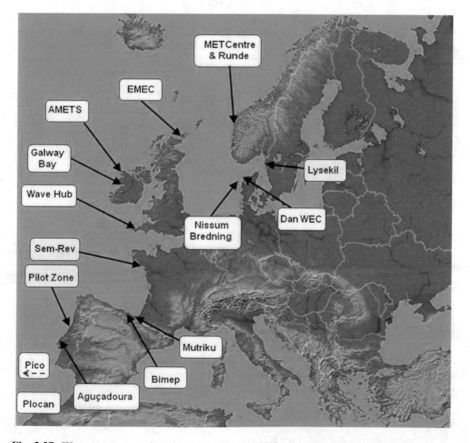

Fig. 2.27 Wave energy test sites throughout Europe [41]

In [21] details, including wave data and more, are given for a number of the illustrated test sites. From the detailed wave data given here, the graphs in Figs. 2.28 and 2.29 have been generated.

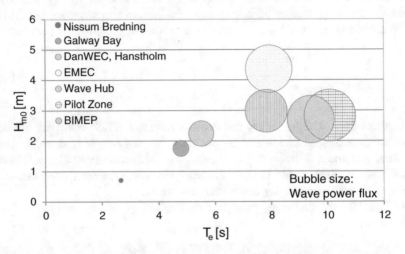

Fig. 2.28 Characteristic power production wave conditions (significant wave height and energy period) at selected European test sites (Weight averaged with contribution to mean wave power flux)

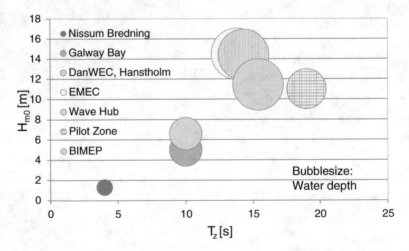

Fig. 2.29 Extreme wave conditions (significant wave height and average period) at selected European test sites

From the figures it is clear to see that the test sites cover a rather wide range of sea states, in production as well as extreme conditions. This corresponds well with the need for real sea test sites from pre-commercial scales to sites with conditions corresponding to the harsh conditions the WECs will be facing at fully commercial sizes.

It is interesting to note that especially the BIMEP and the Pilot Zone sites are dominated by significantly longer waves in production conditions. This has to be carefully considered when designing the WEC prototypes for these locations, as this for most types of WECs means that these sites primarily are well suited for very large devices, as tuning the WEC to larger wave periods inevitably leads to a larger structure.

References

1. IPCC: Climate Change 2014: Synthesis Report. Contribution of Working Groups I, II and III to the Fifth Assessment Report of the Intergovernmental Panel on Climate Change [Core Writing Team, R.K. Pachauri and L.A. Meyer (eds.)]. IPCC, Geneva, Switzerland, 151 pp (2014). http://www.ipcc.ch/pdf/assessment-report/ar5/syr/SYR_AR5_FINAL_full.pdf
2. Hubbert, M.K.: Nuclear Energy and the Fossil Fuels, Publ. no. 95, Shell Development Company, Houston, Texas, June 1956. http://www.hubbertpeak.com/hubbert/1956/1956.pdf
3. "Grøn energi - vejen mod et dansk energisystem uden fossile brændsler" er udgivet af Klimakommissionen. © 2010 Klimakommissionen ISBN: www 978-87-7844-878-1. I den forcliggende udgave er enkelte tal i boks 4.1 på side 81 rettet. Denne publikation er også tilgængelig i en trykt version ISBN: 978-87-7844-879-]
4. Nielsen, K., Rubjerg, M., Hesselbjerg, J.: Kortlægning af bølgeenergiforhold i den danske del af Nordsøen, Energistyrelsen, J. no. 51191/97-0014
5. Isaacs, J.D., Schmitt, W.R: Ocean Energy: Forms and Prospects. Science, New Series, Vol. 207, No. 4428, pp. 265–273 (1980). Published by: American Association for the Advancement of Science, Article Stable. http://www.jstor.org/stable/1684169
6. Assessing the Global Wave Energy Potential, Paper no. OMAE2010-20473, ASME 2010 29th International Conference on Ocean, Offshore and Arctic Engineering (OMAE2010), vol. 3, pp. 447–454. June 6–11, 2010, Shanghai, China. http://dx.doi.org/10.1115/OMAE2010-20473. ISBN 978-0-7918-4911-8
7. http://en.wikipedia.org/wiki/World_energy_consumption. Accessed 03 Aug 2012
8. http://en.wikipedia.org/wiki/Electric_energy_consumption. Accessed 03 Aug 2012
9. EPRI: Mapping and Assessment of the United States Ocean Wave Energy Resource, Report 1024637, Palo Alto, CA, USA (2011)
10. SI Ocean, Wave and Tidal Energy Market Deployment Strategy for Europe (2014). http://www.oceanenergy-europe.eu/images/projects/SI_Ocean_Market_Deployment_Strategy_-_Web_version.pdf
11. TPWind Advisory Council: Wind Energy: A Vision for Europe in 2030 (2006). http://www.windplatform.eu/fileadmin/ewetp_docs/Structure/061003Vision_final.pdf
12. Energistatistik 2010, Energistyrelsen (2011). ISBN 978-87-7844-908-5
13. Kofoed, J.P.: Ressourceopgørelse for bølgekraft i Danmark, DCE Contract Reports, nr. 59, Aalborg Universitet. Institut for Byggeri og Anlæg, Aalborg (2009)
14. http://www.outsidelands.org/wave-tidal.php
15. Salter, S.H.: Wave power. Nature **249**(5459), 720–724 (1974)
16. http://www.ocean-energy-systems.org/ocean_energy/waves/. Accessed 08 March 2012
17. Ingram, D., Smith, G., Bittencourt-Ferreira, C., Smith, H. (eds.): Protocols for the equitable assessment of marine energy converters. University of Edinburgh, School of Engineering, Edinburgh, United Kingdom (2011). ISBN: 978-0-9508920-2-3
18. Hamedni, B., Mathieu, C., Bittencourt-Ferreira, C.: Generic WEC System Breakdown. D5.1 of SDWED project. DNV-GL, Aalborg University (2014). http://www.sdwed.civil.aau.dk/digitalAssets/97/97538_d5.1.pdf

19. http://www.pelamiswave.com/pelamis-technology
20. http://www.equimar.org/
21. Nielsen, K., Pontes, T.: Generic and Site-related Wave Energy Data. Final Technical Report. Annex II—Guidelines for Development and Testing of Ocean Energy Systems. OES-IA Document No: T02-1.1 (2010). http://www.ocean-energy-systems.org/library/oes-reports/annex-ii-reports/document/generic-and-site-related-wave-energy-data-2010-/
22. http://www.ens.dk/en-US/policy/danish-climate-and-energy-policy/danishclimatecommission/greenenergy/Sider/Forside.aspx. Accessed 03 Jan 2012
23. http://www.cnet.com/pictures/renewable-energy-stages-a-comeback-images/3/
24. http://www.voithhydro.com/vh_en_pas_ocean-energy_wave-power-stations.htm
25. http://www.pico-owc.net/
26. http://renews.biz/62747/oceanlinx-in-rescue-mode/
27. http://www.pelamiswave.com/image-library
28. http://www.waveenergyfyn.dk/info%20Waveenergyfyn%20test%20rapport.htm
29. http://www.dexawave.com/photos-for-download.html
30. http://www.oceanpowertechnologies.com/products.htm
31. https://dublin.sciencegallery.com/gm_wavebob
32. http://www.seabased.com/
33. http://www.carnegiewave.com/index.php?url=/ceto/what-is-ceto
34. http://www.flickr.com/photos/aquamarinepower/6424528513/in/photostream/
35. http://aw-energy.com/news-media/gallery
36. http://d2r42o2f7hk334.cloudfront.net/wp-content/uploads/2013/02/rme-wave-energy-converter.jpg
37. http://www.langleewavepower.com/media-centre
38. http://www.wavestarenergy.com
39. http://www.weptos.com/press/pictures/
40. http://www.waveenergy.no/Technology.htm
41. Ref.: http://www.sowfia.eu/index.php?id=3
42. Cruz, J.: Ocean Waves Energy—Current Status and Future Perspectives. Springer Series in Green Energy and Technology (2008). ISSN 1865-3529. ISBN 978-3-540-74894-6
44. http://www.emec.org.uk/marine-energy/wave-devices/. Accessed 15 March 2012

Chapter 3
The Wave Energy Resource

Matt Folley

3.1 Introduction to Ocean Waves

Nobody who has spent time looking at the waves from the beach and seen the damage that they can cause to coastal structures can doubt that they can contain large amounts of energy. However, to convert this wave energy resource into a useable form requires an understanding of its characteristics. Of course, the wave energy industry is not the first industry that has had the need to understand and characterise ocean waves; however, it is important to recognise that the required ocean wave characteristics for wave energy converters are somewhat different from the typical characteristics used for other industries.

3.1.1 Origin of Ocean Waves

A combination of a variety of different disturbing and restoring forces can create the waves on the ocean surface, as shown in Fig. 3.1. Thus, in general terms, tides could be considered as very long period waves and disturbances such as earthquakes that cause tsunamis could also legitimately be called waves. However, the waves that are exploited by wave energy converters are generally generated by wind blowing across the surface of the ocean.

M. Folley (✉)
Queen's University Belfast, Belfast, UK
e-mail: m.folley@qub.ae.uk

© The Author(s) 2017
A. Pecher and J.P. Kofoed (eds.), *Handbook of Ocean Wave Energy*,
Ocean Engineering & Oceanography 7, DOI 10.1007/978-3-319-39889-1_3

43

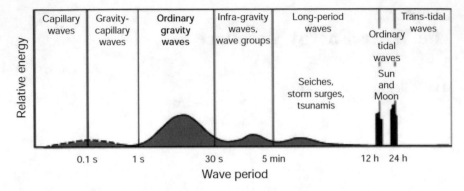

Fig. 3.1 The types of waves that may occur in the ocean [1]

If the process of waves generated by the wind is considered in greater detail, the waves always start as small ripples, but increase in size due to the sustained energy input from the wind. Provided that the wind continues to blow then the waves reach a limit, beyond which they do not grow due to energy losses such as white-capping balancing out the energy input from the wind. In this case the waves are considered to be fully developed. Whether or not a sea is fully developed will depend on both the wind speed and also the distance, or fetch, over which the wind has been blowing. However, when the wind stops blowing the waves will continue to exist and can travel for very large distances with virtually no loss of energy. In this state, they are typically called swell waves because the wind responsible for their

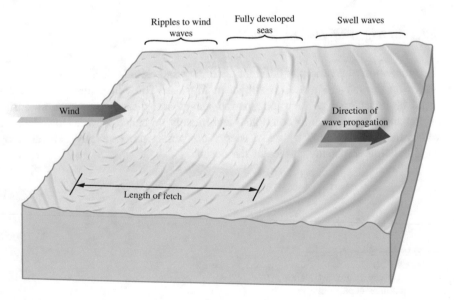

Fig. 3.2 The generation of ocean waves. Adapted from [2]

generation is no longer present. Figure 3.2 shows a simplified representation of these processes.

Using this representation of wave generation, it is common to separate waves into wind waves created by local winds and swell waves, created by to winds that are no longer blowing. Whilst the separation of waves into wind and swell waves may be useful for discussion, it should be recognised that they are essentially two extremes of a continuum of waves. In actuality, all waves are both created by the effect of previous wind and affected by the local wind. Thus, although separation of waves into wind and swell waves may be a useful tool for describing the conditions of a particular location in the ocean, this is simply an abstraction and there is no fundamental difference in the hydrodynamics of wind and swell waves.

3.1.2 Overview of the Global Wave Energy Resource

Figure 3.3 shows the global variation in the annual average omni-directional wave power density. This figure shows that the main areas of wave energy resource occur in bands across the Northern and Southern hemispheres, with less energetic regions close to the equator and poles. However, as discussed in more depth in Sect. 3.3, the annual average omni-directional wave power density does not include significant amounts of information that are vital to determining the utility of a particular wave energy resource for a particular wave energy converter. It is possible to produce a range of other figures that show how other important factors vary across the globe, and some of these factors have been reported by others [3, 4]; however, the specificity of individual wave energy converters means that it is not possible to be highly prescriptive regarding the appropriate important factors. Indeed, as discussed in Sect. 3.3, particular care must be taken when reducing the wave climate

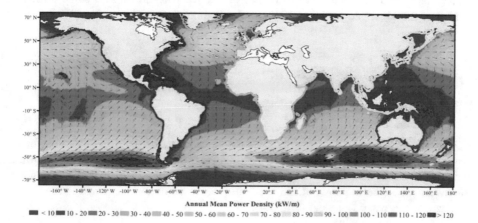

Fig. 3.3 The global wave energy resource [5]

to a finite set of factors since this data reduction always results in a loss of information, which could distort the representation of the resource and thus potentially the conclusions that may be drawn from it.

As an alternative to using factorisations that define the global wave resource, it is possible to use an understanding of the local meteorology and geography to estimate the expected characteristics of the wave resource. Although the results of such an analysis may be more qualitative than quantitative, it also provides a more direct understanding of the conditions and minimises the potential of obtaining a distorted view of the wave resource. Using the understanding of how waves develop, it is possible to state that the most energetic wave climates may be expected to occur where there is a large body of water, with weather systems that track in the same direction as the direction of wave propagation. In these cases, the winds associated with the weather system will cause the waves to continue growing across the whole body of water resulting in the largest waves. Thus, the waves that reach the western coast of Europe are typically larger than those that reach the eastern coast of the USA because not only do the winds normally blow west to east, but the typical track of the weather systems in the North Atlantic is west to east.

Knowledge of the typical wind directions and fetch lengths can also be used to provide an initial indication of the type of waves that may be expected at a particular location. For example, waves in the Mediterranean Sea are typically small because the fetch lengths are also relatively small. Conversely, waves in the South Pacific Ocean are typically large because of the large fetch lengths and relatively high winds in this region, especially in the higher latitudes. Finally, waves in the equatorial regions are typically relatively small because the wind speeds in this region are also typically small.

The seasonal weather variations can also be used to understand the consistency of the wave resource. A significant factor here is that the wind in the Southern hemisphere is significantly more consistent than the winds in the Northern hemisphere so that the wave resource is much less variable. Thus, it is possible to produce reasonable qualitative estimates of the wave resource with some knowledge of the local geography and meteorological conditions.

3.2 Water Wave Mechanics

3.2.1 Definition and Symbols

A basic wave is typically considered as a sinusoidal variation at the water surface elevation and can be defined as having a height, H, which is the vertical distance from the wave crest to the wave trough, a wavelength, λ, which is the distance between two similar points of the wave and the wave period, T, which is the time taken for the wave to repeat (Fig. 3.4).

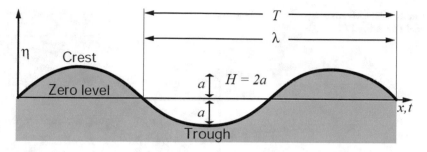

Fig. 3.4 Definition of wave parameters over a sinusoidal wave [1]

In addition it is useful to define a number of other wave parameters

$$\text{Wave steepness}, s = H/\lambda \qquad (3.1)$$

$$\text{Wave number}, k = 2\pi/\lambda \qquad (3.2)$$

$$\text{Wave frequency}, \omega = 2\pi/T \qquad (3.3)$$

Of these additional parameters, the wave steepness is often used to distinguish between linear and non-linear waves. Typically, if the steepness is less than 0.01 then the linear wave relationships are valid, but as the steepness increases then linear theory becomes less accurate and higher-order wave models such as the 5th order Stokes waves are more appropriate [6]. However, in actuality it is very difficult to use the higher-order wave models for analysing anything other than regular waves and so linear wave theory is often used for waves much steeper than 0.01.

Figure 3.5 shows the suitability of the different wave theories based on the wave steepness $s = H/gT^2$ and the relative water depth h/gT^2. Although useful to note the applicability of different wave theories, it is worth noting that it is extremely complex to apply any theory, except linear theory, to irregular waves. Consequently, it is common to use linear wave theory beyond the bounds shown in Fig. 3.5, but recognising that it is not entirely correct.

3.2.2   Dispersion Relationship

An important characteristic of ocean waves is that they are generally dispersive, which means that the energy in the wave does not travel at the same velocity as the wave profile [6]. The effect of dispersion can be seen when a stone is dropped into water or the wave paddles in a wave tank stop generating waves. In this case waves appear to be left behind the main wave and are travelling at a slower velocity than the wave crests due to the wave energy. The velocity of a wave crest is typically

Fig. 3.5 Chart of wave
model suitability [7]

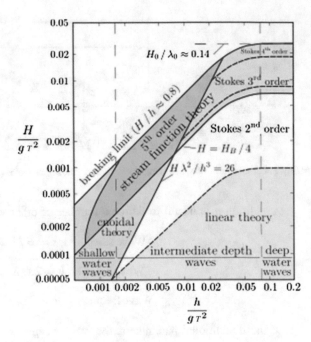

called the wave celerity, c, and the velocity of the energy propagation is typically
called the group velocity, C_g. In deep water the group velocity is equal to a half of
the wave celerity, but in general the group velocity is given by

$$C_g = \frac{1}{2}\left[1 + \frac{4\pi d/\lambda}{\sinh(4\pi d/\lambda)}\right]c \qquad (3.4)$$

Moreover, not only does the group velocity vary with water depth, but the wave
celerity also varies with water depth and is given by

$$c = \frac{\lambda}{T} = \frac{gT}{2\pi}\tanh\left(\frac{2\pi d}{\lambda}\right) \qquad (3.5)$$

This is called the dispersion equation and defines the wavelength based on the
wave period and water depth.

3.2.3 Water Particle Path and Wave Motions

In this case the water surface elevation ζ is given by

$$\zeta = \frac{H}{2}\cos\left[2\pi\left(\frac{x}{\lambda} - \frac{t}{T}\right)\right] \qquad (3.6)$$

However, this variation in water surface elevation is actually the result of an elliptical motion of the water particles, which also extends far below the water surface, with the amplitude of motion decreasing exponentially with depth as shown in Fig. 3.6. Thus, the vertical displacement of the water particles $\zeta(z)$ is given by

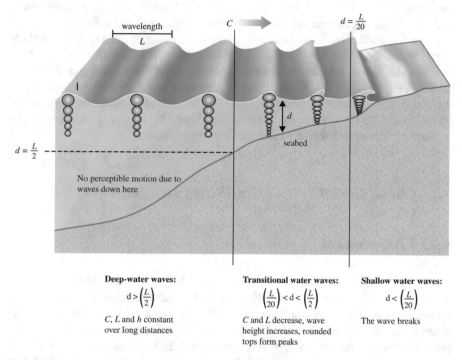

Deep-water waves:

$$d > \left(\frac{L}{2}\right)$$

C, L and h constant over long distances

Transitional water waves:

$$\left(\frac{L}{20}\right) < d < \left(\frac{L}{2}\right)$$

C and L decrease, wave height increases, rounded tops form peaks

Shallow water waves:

$$d < \left(\frac{L}{20}\right)$$

The wave breaks

Fig. 3.6 Motion of water particles in deep and shallow water. Adapted from [2]

$$\zeta(z) = \frac{H}{2}\cos\left[2\pi\left(\frac{x}{\lambda} - \frac{t}{T}\right)\right]\frac{\sinh[2\pi(z+d)/\lambda]}{\sinh[2\pi d/\lambda]} \qquad (3.7)$$

and the horizontal displacement $\xi(z)$ is given by

$$\xi(z) = -\frac{H}{2}\sin\left[2\pi\left(\frac{x}{\lambda} - \frac{t}{T}\right)\right]\frac{\cosh[2\pi(z+d)/\lambda]}{\sinh[2\pi d/\lambda]} \qquad (3.8)$$

Thus, in deep water the water particle motions are circular, but they become more elliptical as the water depth decreases as shown in the Fig. 3.6.

In particular, it can be seen that the variation in water particle motion is dependent on the water depth relative to the wavelength, and this is often used to define three regions of water depth: 1) deep water where the seabed does not affect the waves and typically requires the water depth to be greater than half the wavelength, 2) shallow water where there is no variation in horizontal water particle motion with water depth and typically requires the water depth to be less than 1/20th of a wavelength and 3) intermediate depth that exists between these two extremes [6].

At a depth of half a wavelength, the wave-induced motions are only approximately 4 % of those at the surface and thus could be considered insignificant. However, it should always be remembered that these limits are somewhat arbitrary, and since they depend on the wavelength this means that the definition of water depth is not fixed. That is, a site may be defined as being in deep water for a short wave, whilst the same site for a different wave may be in intermediate water. Thus, care should always be taken to determine which reference wavelength should be used to define the relative depth. For wave energy, it is particularly important to recognise this condition, because many wave energy converters defined as "deep-water" devices, such as Pelamis, are typically deployed in what many oceanographers would define as intermediate water depths.

3.3 Characterisation of Ocean Waves and the Wave Climate

3.3.1 Introduction

Traditionally, sea-states have been characterised using a representative wave height, which before any method for recording waves existed was based solely on observation. That is, the representative wave height was defined as the wave height as reported by an "experienced observer", whom we must assume had spent many years listening to the estimates of other experienced observers so that a relatively consistent estimate of the wave height could be made. This was called the Significant Wave Height, symbolized by H_s. However, it is clear that the accuracy of this method is highly dependent on the experience of the observer and as such subject to significant error. When it became possible to record the variation in the water surface elevation, an alternative method of defining wave height was developed. With a record of the variation in the water surface elevation, it is possible to measure the height of individual waves and thus produce a more reliable estimate of the wave height. In order to be consistent with historical reports, it was decided that the new records of surface elevation should be analysed so as to produce an estimate equivalent to the H_s. It was found that a good estimate of H_s was given using the average height of the third highest waves.

In modern times, the variation in wave surface elevation is typically recorded digitally, which provides the potential for significant analysis of the wave record. The most significant development in the representation of the sea is the definition of the sea using a spectrum. To understand the concept of the wave spectrum, it is first necessary to accept that the variation in water surface can be represented as the linear super-position of sinusoidal waves of different frequencies, amplitudes, directions and phases. Although this representation could be viewed as simply a change in the co-ordinate system (from the time-domain to the frequency-domain), it actually appears to be a reasonably good representation of the underlying physics. Indeed, the wave spectrum is now generally used to fully define any sea-state, with the assumption that there is a random phase between all of the individual wave components, which is a natural consequence of the assumption of linear super-position.

Although the linear super-position and random phase assumptions are commonly applied, it is important to recognise that they are not universally valid. The most obvious example of the breakdown of linear super-positioning is in sea-states with steep and breaking waves. The super-positioning of the waves means that the waves will become unstable and break at particular times and locations, resulting in a variation in the water surface elevation that is poorly represented by a random phase, linear super-position of the wave components. Another case where linear super-positioning does not provide an accurate representation of the water surface elevation is in shallow water, where a single frequency wave would be more accurately represented by a cnoidal or higher-order Stokes wave than by a sinusoidal wave. Notwithstanding these cases, it is still typical to use a spectrum and linear super-positioning to represent the sea-state because no reasonable alternative currently exists. However, it is important to recognise that, in general, this representation of the sea is, to some extent, simply an abstraction that has been found to be very useful in defining the sea.

3.3.2 Temporal, Directional and Spectral Characteristics of the Wave Climate

To understand how the ocean waves may influence the performance of a wave energy converter, it is useful to consider the temporal, directional and spectral characteristics of the ocean waves and how these may influence the relationship between the average omni-directional wave power and the average power generation.

Firstly, the **temporal characteristic of a wave climate** is how the sea-states that make up a wave climate vary in time as illustrated in Fig. 3.7 for the significant wave height. In general, the more consistent the wave climate is, the more attractive it becomes (for a particular average wave power) because the WEC and power generating plant can remain closest to its optimal operating conditions and thus maximise the system efficiency. However, the sea-states will vary due to changes in the metrological conditions that generate the winds and associated waves. Not surprisingly, the stability of metrological conditions varies across the world so that the wave climate is more consistent in some locations than others. This variability may be primarily associated with daily, seasonal and/or annual variations in the sea-states, each of which will have a slightly different impact on the power generation and its utility. Thus, it is clear that for all locations the temporal characteristics are an important element of the wave climate, which can result in different power generations for the same average incident wave power.

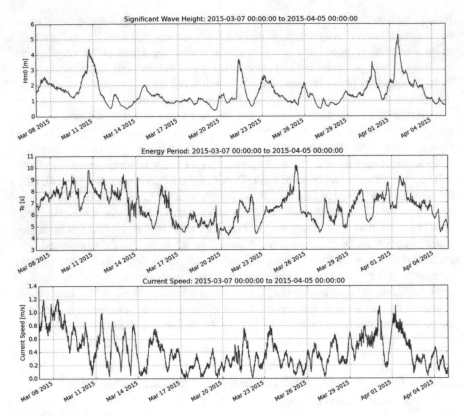

Fig. 3.7 Example of the variation of the significant wave height, energy wave period and tidal current over one month at DanWEC

Secondly, the **directional characteristics of a wave climate** are associated with not only the directional spreading of individual sea-states, but also the directional variation of all the sea-states. The only situation where this may not be critical is for an isolated omni-directional WEC such as a heaving buoy. However, as soon as this buoy is put into a wave farm, the directional characteristics of the sea-state become significant, with the significance increasing with the size of the wave farm. In general, an increase in the directional variation of the wave climate will lead to a reduction in the average power generation because the WECs and/or wave farm will typically be less optimally aligned. As would be expected, the directional characteristics of a wave climate is dependent on its location, which defines the range of weather systems that produce winds, and subsequently waves that contribute to the local wave climate. Thus, the directional characteristics are an important element of the wave climate that should be considered when assessing a potential site.

Finally, the spectral **characteristics of a wave climate** are associated with the wave spectrum of individual sea-states, together with the spectral variation for all the sea-states as illustrated in Fig. 3.8 for the average directional spectra at EMEC. The spectral characteristics of the wave climate can be particularly important because the "efficiency"

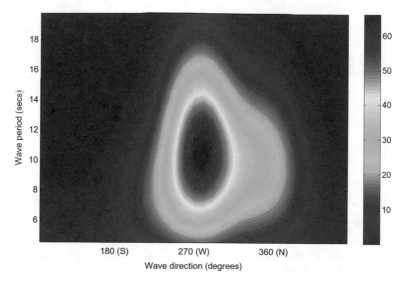

Fig. 3.8 Average directional spectral variance density (m²/Hz/rad) at the European Marine Energy Centre, Orkney, Scotland

of many WECs is frequency-dependent. Thus, wave power associated with particular wave frequencies may be more significant than wave power at other frequencies. For example, the power capture per unit wave height of many surging/pitching devices tends to increase with wave frequency because the incident wave force/torque also tends to increase with wave frequency [8, 9], whilst the incident wave power decreases with wave frequency (see Eq. 3.18). Thus, in this case, incident wave power is clearly not a good proxy for power generation. Thus, in assessing a potential site, it is important consider the spectral characteristics of a potential site, especially in relation to the spectral response of the WECs being considered for deployment at the site.

When a particular site of interest has been identified, then a scatter diagram is often used to characterise it. Figure 3.9 shows an example of a scatter diagram which consists of a frequency of occurrence table indexed by a representative wave period, typically a peak period, zero-crossing period or energy period and a representative wave height, almost always the Significant Wave Height. The scatter diagram clearly provides significantly more information about the wave climate than the average omni-directional wave power; however, it is not without issues. Firstly, depending on the table resolution, the sea-states can vary significantly within any single cell in the scatter table, especially for the cells indexed by a small significant wave height. For example, there is a potential 4:1 variation in wave power between sea-states in a cell defined by Significant Wave Heights of between 0.5 and 1.0 m. Of course, any contribution from these sea-states to either the average incident wave power or power generation is likely to be small and so the impact on any estimate of WEC performance is also likely to be small; however, this may not always be the case and any potential distortion should be considered when using scatter diagrams. Secondly, typically there are no details on the temporal/ directional distribution or spectral shape of the sea-states that are all contained in a single cell, both of

Scatter diagram

Hs \ Tz	3.5	4.5	5.5	6.5	7.5	8.5	9.5	10.5	11.5	12.5	13.5	14.5
0.25	0.0066	0.0056	0.0030	0.0023	0.0011	0.0007	0.0003	0.00005				
1	0.0453	0.1650	0.0906	0.0347	0.0131	0.0047	0.0019	0.00069	0.0001	0.00004	0.00007	0.00005
2	0.0018	0.0368	0.1604	0.0650	0.0229	0.0099	0.0032	0.00121	0.00009	0.00005	0.00005	
3		0.0003	0.0187	0.1084	0.0335	0.0071	0.0033	0.00171	0.0004	0.00007		0.00002
4			0	0.01021	0.05565	0.01163	0.00209	0.00052	0.00034	0.00021	0.00005	
5				0.00002	0.00729	0.02391	0.00301	0.00069	0.00031	0.00014	0.00005	0.00005
6					0.00012	0.00603	0.00691	0.00052	0.00007			
7				0.00002	0.00009	0.00026	0.00352	0.00152	0.00016	0.00005		
8							0.00062	0.00288	0.00017			
9								0.00086	0.00073	0.00002		
10								0.00002	0.00043	0.00016		
11									0.00011	0.00014		
12										0.00004		

Fig. 3.9 Example of a scatter diagram

which can have a significant impact on the power generation of a WEC as discussed above. This second issue is sometimes reduced by producing multiple scatter diagrams for a particular site, which are used to separate the wave climate by peak wave direction or season, but there is clearly a practical limit to the number and range of scatter diagrams that can be produced and used effectively.

Another representation of the wave climate that is often used is the **wave rose** as shown in Fig. 3.10. A wave rose is a graphical representation of the average wave power or Significant Wave Height from different directional sectors. Similar to a set of wave roses may be produced based on season in order to provide additional information that may be useful in understanding the wave climate, especially where different meteorological conditions are responsible for different wave conditions at different times of the year.

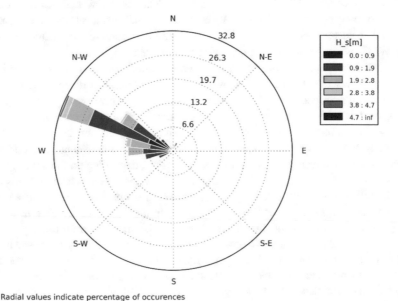

Radial values indicate percentage of occurences

Fig. 3.10 Example wave rose of significant wave height

In summary, it can be seen that the characterisation of the wave climate using single parameters such as the average omni-directional wave power, scatter diagrams and wave roses present only a partial picture of the wave climate. Moreover, care must be taken in translating this partial picture into an estimate of the power generation of a WEC. Indeed, whenever possible it is recommended to use the full time series of directional wave spectra to estimate the average power generation of a WEC [10], and if this is not possible, either because the full data set is not available or it would require too much effort, then it is important to recognise that not only is there an increased uncertainty in the estimate of power generation, but relative performance of the WEC at different locations may also not be a simple function of the average omni-directional wave power or other parameterised characteristic of the wave climate. However, although the use of wave climate characterisations for estimating power generation is not recommended, they do provide an overview that can be useful for understanding the performance of a WEC. Furthermore, as understanding of a WEC increases with identification of the most appropriate characterisations of the wave climate for the particular WEC, it is possible that the WEC's performance could be reasonably estimated from the wave climate characteristics. However, until that point is reached it remains prudent to recognise the limitations of any characterisation of the wave climate and the potential distortion in the estimate of a WEC's power that it may cause.

3.3.3   Spectral Representation of Ocean Waves

In the preceding sections, the wave spectra has been discussed without any clear definition of exactly what it represents. Representation of the ocean using a wave spectrum assumes that it is possible to represent the water surface as the sum of sinusoidal waves with a range of frequencies, amplitudes and directions. The variation of the wave energy with frequency (and direction) is called the wave spectrum. Figure 3.11 shows an illustration of this super-positioning of waves, together with an example of a typical wave spectrum.

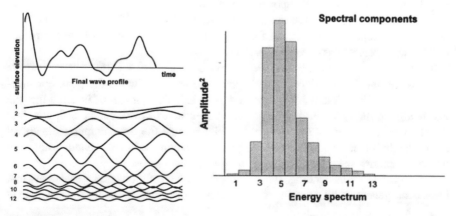

Fig. 3.11 Super-positioning of waves (corresponding to spectral components) to create water surface elevation (*left*) and the resulting spectrum (*right*) [11]

A variety of idealised spectra have been suggested to represent a fully developed sea-state. Perhaps the most commonly used spectrum was developed by Pierson and Moscowitz in 1964 and is called the Pierson-Moscowitz (PM) Spectrum [12]. This spectrum assumes that the wind has been blowing across a sufficiently large expanse of water for sufficiently long that the waves are in equilibrium with the wind, i.e. the sea-state is fully developed and so that the spectrum is dependent on only the wind speed.

Subsequent research completed by Hasselman et al. in the Joint North Sea Wave Observation Project (JONSWAP) identified a refinement to this wave spectrum for when the sea is not fully developed and is based on the wind speed and fetch length [13]. This spectrum is called the JONSWAP spectrum and is commonly used to represent the sea-state that is not fully-developed.

$$S(\omega) = \frac{\alpha g^2}{\omega^5} \exp\left[-\beta\left(\frac{\omega_p}{\omega}\right)^4\right] \tag{3.9}$$

$$S(\omega) = \frac{\alpha g^2}{\omega^5} \exp\left[-\frac{5}{4}\left(\frac{\omega_p}{\omega}\right)^4\right] \gamma^{\exp\left[-\frac{(\omega-\omega_p)^2}{2\sigma^2\omega_p^2}\right]} \tag{3.10}$$

PM spectrum (Eq. 3.9)	JONSWAP spectrum (Eq. 3.10)
$\alpha = 0.0081$	$\alpha = 0.076\left(\frac{U_{10}^2}{Fg}\right)^{0.22}$
$\beta = 0.74$	$\omega_p = 22\left(\frac{g^2}{U_{10}F}\right)^{1/3}$
$\omega_p = \frac{g}{U_{19.5}}$	$\gamma = 3.3$
	$\sigma = \begin{cases} 0.07 \omega \leq \omega_p \\ 0.09 \omega > \omega_p \end{cases}$

where $S(\omega)$ is the spectral variance density, ω_p is the peak frequency, g is the gravitational acceleration, U_{10} is the wind speed at a height of 10 m, F is the fetch length and ω is the wave component frequency.

In addition to the wind speed and fetch length, the JONSWAP spectrum is also defined by the peak enhancement factor γ. This parameter defines how the peakiness of the spectrum as seen in Fig. 3.11.

Comparison of Eqs. 3.9 and 3.10 reveals that the spectral shapes of the JONSWAP and Pierson-Moscowitz spectra are identical when the peak enhancement factor of the JONSWAP spectrum equals 1.0. Thus, it can be inferred that the bandwidth of the spectrum is dependent on its state of development with new and developing seas having a narrower bandwidth, so that the wave components are all at similar frequencies and fully-developed seas having a broader bandwidth, with the wave energy spread over a larger range of frequencies.

To facilitate understanding, the discussion above only considers sea-states that have been generated by a single source of wind. However, in reality the sea-state at a single location may have waves generated from a number of different sources of winds from different directions with different speeds and fetch lengths. Where there are two distinct sources of waves then the sea-state is called bi-modal and has two peaks with different peak directions and frequencies. Figure 3.12 shows an example of a bimodal sea-state. Cases where there are more than two sources of wind result are called multi-modal sea-states. Although there will be some interaction between the waves from the different sources, typically this interaction is small and the spectra can generally be linearly superimposed without too much loss of accuracy (at least when they are not close to breaking).

Fig. 3.12 Example spectrum of a bimodal sea-state

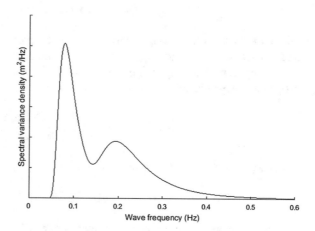

3.3.4   Characterization Parameters

Recalling the introduction to this section, it was noted that there are three different definitions of the Significant Wave Height: the first based on observation, the second based on analysis of a record of the surface elevation time-series and the third based on the wave spectrum. It is important to be aware of which one is being used. The first method is never used nowadays; however, wave data from both the other methods is still commonly used. Thus, it is good practice when referencing the Significant Wave Height to use a subscript to identify the method, with the subscript '1/3' used when the Significant Wave Height is based on the average height of the third highest waves and the subscript 'm0' when the Significant Wave Height is based on the wave spectrum. Unfortunately, in many cases the Significant Wave Height is identified by the subscript 's' and the method used to generate it is unknown. As noted above, the difference between the methods in deep water for a moderate sea-state is relatively small, typically about 1 %; however, the difference increases progressively as the waves steepen and/or water depth decreases.

   In wave energy, the preferred representative wave height is the Significant Wave Height derived from the spectral moments of the wave spectrum, H_{m0}:

$$H_{m0} = 4\sqrt{\int_0^\infty S(\omega)d\omega} \tag{3.11}$$

This is because it is effectively based on the energy in the waves and as such is directly related to the average wave power density. To show this, it is necessary to recognise that with linear super-position the power in each wave can be considered independently and then summed together to give the total average wave power density. Thus, consider a single wave component then the wave power is given by

$$J(\omega) = \rho g S(\omega) \cdot C_g(\omega) \tag{3.12}$$

where the first half of the right-hand side of the equation is the energy in the wave and the second half of the right-hand side of the equation is the velocity at which the energy is propagating, the group velocity [6]. The speed that the wave energy propagates depends on the wave frequency ω and water depth h and is given by

$$C_g(\omega) = \frac{1}{2}\frac{\omega}{k(\omega)}\left(1 + \frac{2k(\omega)h}{\sinh 2k(\omega)h}\right) \tag{3.13}$$

where $k(\omega)$ is defined by the dispersion equation

$$\omega^2 = gk(\omega)\tanh k(\omega)h \tag{3.14}$$

Using the assumption of linear super-positioning, the average wave power density for the sea-state is given by the integral of the wave components

$$J = \int_0^\infty \rho g S(\omega) \cdot \frac{1}{2}\frac{\omega}{k(\omega)}\left(1 + \frac{2k(\omega)h}{\sinh 2k(\omega)h}\right)d\omega \tag{3.15}$$

To progress further, it is useful to define the moments of the spectrum m_n as

$$m_n = \int_0^\infty S(\omega)\omega^n d\omega \tag{3.16}$$

Then, the wave energy period T_e can be defined as the ratio of the first negative moment of the spectrum to the zeroth moment of the spectrum as given by Eq. 3.17

$$T_e = \frac{m_{-1}}{m_0} \tag{3.17}$$

and the Significant Wave Height ($H_{m0} = 4\sqrt{m_0}$, see Eq. 3.11) can be used directly in the calculation of the wave power density. Consequently, the omnidirectional wave power J can be defined in deep water as

$$J = \frac{\rho g^2}{64\pi} H_{m0}^2 T_e \tag{3.18}$$

In addition to defining the Significant Wave Height and Energy Period, the moments of the spectrum can also be usefully used to define other characteristics of a sea-state. For example, the relative spreading of the energy with wave frequency, often termed the spectral bandwidth ϵ_0, can be defined as the standard deviation of the period variance density, normalized by the energy period, which is

$$\epsilon_0 = \sqrt{\frac{m_0 m_{-2}}{m_{-1}^2} - 1} \tag{3.19}$$

In addition, it is possible to make a spectral estimate of the mean zero-crossing period of the waves T_z, which is given by

$$T_z \cong T_{02} = \sqrt{\frac{m_0}{m_2}} \tag{3.20}$$

This spectral estimate of the mean zero-crossing period of a sea-state is useful because it allows a spectrum to be scaled using assumptions regarding the spectral shape and the mean zero-crossing period, which is a common parameter used to define historical wave resource data. Similarly, the spectrum can be scaled using the peak period T_p, which has also been commonly used to define wave resource data.

Using these expressions, it is also possible to calculate the ratio between different measures of the wave period for particular spectral shapes. This can be especially useful when it is considered necessary to convert between representations of the wave period. For example, for a JONSWAP spectrum with a peak enhancement factor, $\gamma = 3.3$, the ratios of the wave periods are

$$1.12\, T_e = 1.29\, T_z = T_p \tag{3.21}$$

For many devices, and for all wave farms, the directional characteristics of the sea-state will also be important. The directionally resolved wave power density $J(\theta)$ is a key directional characteristic of the sea-state as it defines the wave power propagation in a particular direction. The directional wave spectrum can be used to calculate the variation in the directionally resolved wave power density $J(\theta)$ as given by

$$J(\theta) = \rho g \int_{-\pi}^{+\pi} \int_0^{\infty} S(\omega, \varphi) C_g(\omega) \cos(\theta - \varphi) \delta \cdot d\omega \cdot d\varphi$$
$$\begin{cases} \delta = 1, & \cos(\theta - \varphi) \geq 0 \\ \delta = 0, & \cos(\theta - \varphi) < 0 \end{cases} \tag{3.22}$$

Other direction parameters that can be derived from this and provide additional characteristics of a sea-state include the direction of maximum directionally resolved wave power density and the directionality coefficient, which is the ratio of the maximum directionally resolved wave power density and the omni-directional wave power density as defined in Eq. 3.15.

A further characterisation of the waves, which may be particularly important when considering transient effects, is the tendency for larger waves to be grouped together; this characteristic is called wave groupiness. Non-linear processes can play an important part in the creation of wave groups, especially in shallow water; however, it is also dependent on the spectral bandwidth, with narrow-banded spectra generally having higher levels of wave groupiness than broad-banded spectra. A common measure of wave groupiness is the *average run length*, which is the average number of consecutive waves that exceed a specified threshold such as the significant or mean wave height.

The characterisations of a sea-state described above have generally proved useful for analysing the wave climate for wave energy converters; however, it must always be appreciated that all these characterisations reduce the amount of information available on the wave climate. When available, it is almost always better to use the directional wave spectrum of a sea-state in any analysis rather than a characterisation of the sea-state. Unfortunately, this is not always possible either because the directional wave spectrum is not available or because it would require excessive amounts of computational effort. However, whenever only the sea-state characteristics are presented, or used in an analysis, it is important to be cautious with any conclusions because of the potential distortions that can occur.

3.3.5 Challenges in Wave Climate Characterisation

A wave climate can be reasonably approximated as a long-term series of sea-states that are defined by the directional wave spectrum. Together with other metocean parameters such as water depth, marine current speed/direction and wind speed/direction, this can be used to estimate the power capture and design parameters for any wave energy converter deployed at the location. However, typically it is not possible to work with this amount of data (or the data is not available) and so a characterisation of the wave climate is used. The wave climate characterisation can essentially be one of two types: the characterisation of the **wave climate at a single point or the characterisation of the wave climate over an area**. However, it is important to recognise that in either case the characterisation results in a compression of the details on the wave climate and so does not contain all the information that may be relevant to the performance of a wave energy converter.

The **average omni-directional wave power** is probably the most common characterisation of the wave resource for the assessment of wave energy. This seems, and likely is, a reasonable characterisation since it is clear that to extract significant amounts of wave energy the incident wave power must also be significant; without waves there is no wave power. Figure 3.3 shows an example of this characterisation illustrating the variation in the average omni-directional wave power around the world. As would be expected, the areas with higher average omni-directional wave power, such as the north-west coast of Europe, are also the areas with the most interest in the deployment of wave energy converters. The implicit assumption is that a wave energy converter's power capture is proportional to the average omni-directional wave power thus a larger average omni-directional wave power equates to a larger power capture. However, whilst it may be reasonable to assume that a wave energy converter will produce more power at a site with an average omni-directional wave power of 40 kW/m compared to an alternative site with 2 kW/m, it is less clear that this will be true if the alternative site had an average omni-directional wave power of 30 kW/m.

The key factor to consider is that when comparing potential sites the use of the average omni-directional wave power obscures information regarding the temporal, directional and spectral characteristics of the wave climate (see Sect. 3.2) that may be important to the average power capture. Of course, how these characteristics may affect the average power generation will vary with the WEC and so it is difficult to be overly prescriptive regarding the extent of distortion that may be due to using average omni-directional wave power as a proxy for average power generation. One method to compensate for the potential distortion is to provide information on other aspects of the wave climate simultaneously with the average omni-directional wave power. Examples of this additional information could include the ratio of maximum wave power to average wave power, the average directionality coefficient, the average spectral width, and/or the average energy period. Unfortunately, whilst this additional information does provide more details of the characteristics of the wave resource that may suggest the relative strengths and weaknesses of particular sites, it still does not provide a clear indication of how a WEC's power generation may differ between locations.

Whilst it is frustrating that a single parameter, or even set of parameters, cannot be used to assess the suitability of a potential WEC deployment site, this is the state of the wave energy industry at the moment. The rich diversity of WEC concepts currently being developed means that there are a multiple of relationships between the wave resource and power generation. Moreover, it is possible that a particular WEC concept may be most suitable at one location, whilst another WEC concept is more suitable at another location. Thus, there may not be the complete convergence onto a single concept as in wind energy, with the three-bladed horizontal-axis turbine, due to the potentially greater diversity of wave resource characteristics compared to wind resource characteristics, which is generally successfully characterised simply by the average wind speed.

Although not associated with a particular WEC concept, a useful illustration of the dangers of using the average omni-directional wave power as a proxy for power generation is in assessing the effect of water depth on the incident wave power. Off of the west coast of Scotland, the average omni-directional wave power decreases as it approaches the shore and the water depth reduces, so that in 10–20 m of water it is only typically 50 % of its offshore value. To assess the extent that this reduction in average omni-directional wave power may translate to a reduction in potential power generation, it is necessary to consider how the change in average omni-directional wave power has occurred. Consideration of the wave propagation process indicates that there are **six main processes responsible for the change in average omni-directional wave power**, namely: shoaling, refraction, diffraction, depth-induced wave breaking, bottom friction and wind growth, which are each considered in detail below.

3.3.6 Coastal Processes

3.3.6.1 Shoaling

Shoaling can be understood by considering a wave propagating into shallower water. When a wave propagates into shallower water, the wave group velocity changes, but the change in group velocity is not accompanied by a change in energy flux. Thus, conservation of energy means that the wave height must get larger in order to keep the total energy flux constant. It can be visualised as a bunching up of the incident waves so that they increase in height as illustrated in Fig. 3.13.

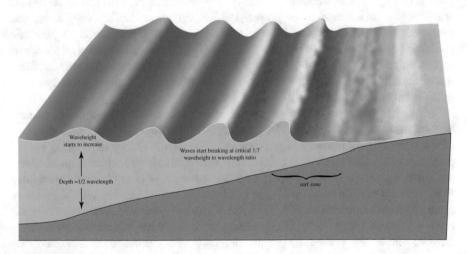

Fig. 3.13 Change in wave shape due to water depth. Adapted from [7]

3.3.6.2   Refraction

To understand refraction, consider a wave propagating at an angle to the depth contours. In this case, the dispersion equation tells us that the part of the wave crest in shallow water will travel slower resulting in a turning of the direction of wave propagation. This effect explains why on the beach all the waves appear to come from a direction approximately orthogonal to the coastline (Fig. 3.14).

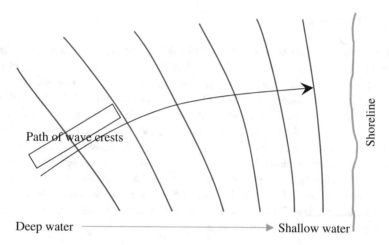

Fig. 3.14 Wave refraction near the shore due to a change in water depth

Refraction causes the waves to change direction so that their propagation direction is more orthogonal to the seabed depth contours. The net effect of this on all the waves is to reduce the directional spreading in the waves so that as the water depth reduces their approach from a more concentrated direction. In addition, any single wave component refraction causes a reduction in the wave height as it is spread out over a larger distance; however, it is important to remember that the refraction process is energy conserving and thus does not change the amount of energy travelling orthogonally to the depth contours. Using the west coast of Orkney again as an indicative site, it can be seen that refraction causes a significant change in the average omni-directional wave power between an offshore and nearshore site, with a reduction of 30–40 %. Thus, the key to assessing whether or not the change in average omni-directional wave power due to refraction will similarly affect the power generation depends on the directional sensitivity of the WEC or wave farm. An isolated heaving buoy may be insensitive to wave direction and so the reduction in average omni-directional wave power will result in a similar reduction in power generation; however, a wave farm will have a directional sensitivity so that the incident wave power is defined by the wave power incident on the wave farm. If this is the case and the wave farm is aligned with the depth contours, then refraction would have no effect on the relevant incident wave power as refraction is energy conserving. In reality, it is likely that the wave farm will not be aligned with the depth contours and so refraction will change the power incident on the wave farm, with the effect due to refraction increasing as the angle between the wave farm and depth contours increases. Thus, again it can be seen that the suitability of using average omni-directional wave power as a proxy for power generation depends on the WEC characteristics and deployment configuration.

3.3.6.3 Diffraction

Diffraction occurs when waves meet a surface-piercing obstacle such as an island, headland or breakwater. Without diffraction the waves would continue to travel in the same direction leaving a region of calm water in the lee of the obstacle. However, diffraction means that the waves will bend so that there are waves behind the obstacle. The amount of diffraction depends on the wavelength, with the longer waves diffracting to a greater extent than the shorter waves. If there is more than one source of diffraction, e.g. either side of an island, then a diffraction pattern may form where there are areas of increased and decreased wave height due to constructive and destructive interference. Although diffraction means that waves will occur on the leeward side of an obstacle, generally these waves will be smaller than the incident waves (except in the special case of constructive interference) so the wave resource behind an obstacle, is likely to be smaller than the seaward wave resource.

3.3.6.4 Depth-Induced Wave Breaking

Wave breaking occurs when the horizontal wave particle velocity becomes greater than the wave celerity. When this occurs the wave will spill energy in the form of breaking waves. Depth-induced wave breaking is related to the steepening of the waves in shallow water due to shoaling. When the wave height is greater than about 0.8 of the water depth (or about 0.14 of the wavelength), then the waves break. There are three different types of breaking waves: spilling, plunging and surging as shown in Fig. 3.15, depending on the wave and seabed steepnesses (or more specifically the Iribarren number [14]).

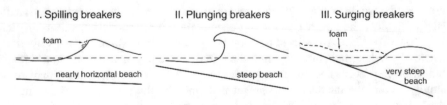

Fig. 3.15 Classification of breaking waves [15]

In water depths greater than about 10 m, the vast majority of waves will not break and so it is tempting to consider that this process is not significant in assessing the suitability of using average omni-directional wave power to compare offshore and nearshore sites. However, the average omni-directional wave power includes energy from all events irrespective of its exploitability. In particular, it includes the wave energy in storms, which at the offshore site, in deep water, may have 40–50 times the wave power of the average wave power. Thus, although storms may only occur infrequently, they may make a relatively large contribution to the average wave power and account for perhaps 15–20 % of the total wave energy. On the contrary, at the nearshore site the wave energy in a storm is a much smaller multiple of the average wave power because depth-induced wave breaking has limited the wave energy in a storm that reaches the nearshore, but not affected the wave power in the most commonly occurring seas. The proportion of the total wave energy contained in storms is important since it is largely un-exploitable, either because the WEC power generation is limited by the plant rating, or because it has to shut down in order to survive the storm. Thus, because the average omni-directional wave power does not distinguish whether or not the wave energy is exploitable, it distorts the relative potential power generation at the offshore and nearshore sites.

3.3.6.5 Bottom Friction

The reduction of average omni-directional wave power has often been primarily attributed to bottom friction. However, as illustrated above, a significant proportion of the reduction is caused by other factors and in particular refraction [16]. Indeed, for a typical seabed bottom friction only accounts for about 5 % of the reduction in average omni-directional wave power. The reduction in spectral wave energy due to bottom friction is complex and varies with depth so that the wave spectrum changes as a result of bottom friction, although the small amount of energy reduction means that the change in spectrum will also be small. However, as different WEC concepts have different spectral responses, it is possible that the change in spectral shape will be more significant for one WEC concept than another. Thus, it is possible that the change in average omni-directional wave power due to bottom friction has a different impact on average power generation for different devices because of their different response characteristics.

3.3.6.6 Wind Growth

As there is a larger fetch to the open ocean for the nearshore, it may be expected that wind growth will increase the wave power at this site. Unfortunately, in many cases the offshore waves are already in equilibrium with the wind because of the large fetch and so they cannot grow significantly between offshore and nearshore. However, when the wind blows from the land there will be minimal fetch for the nearshore site, but the fetch may be a significant for the offshore site. At a location 40 km off of the west coast of Orkney, Scotland, the waves travelling away from the shore typically account for about 15 % of the average omni-directional wave power, which is not an insignificant proportion of the total wave power. The key to assessing whether this 15 % of wave power should or should not be included depends on whether or not the WEC can capture energy travelling in the opposite direction to the majority of the waves. For example, a heaving buoy, such as the Wavebob, may be expected to exploit this wave power, but an overtopping device, such as the WaveDragon, is less likely to be able to exploit it so that the omni-directional wave power is less appropriate as a proxy for power generation in this case.

3.3.7 Case Study—Incident Wave Power

Figure 3.16 shows an example of the effect of considering these factors for the wave climate at the European Marine Energy Centre (EMEC), where gross power refers to the omni-directional wave power, net power includes the effect of wave refraction and exploitable power also includes the effect of wave breaking and bottom friction. It can be seen that the difference in exploitable resource from a 50 m "deep water" site to a 10 m "shallow water" site is around 20 %, significantly less than the 50 % reduction in resource that can arise from an inappropriate use of the omni-directional wave power. More details on this case study are available in [17].

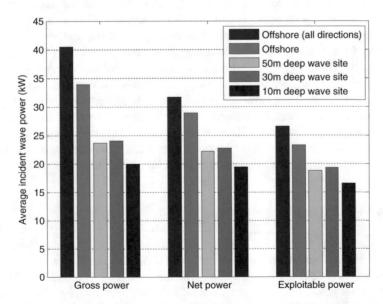

Fig. 3.16 Average incident wave power at the European Marine Energy Centre, Orkney, Scotland [17]

The analysis and discussion above illustrate that the change in average omni-directional wave power from offshore to nearshore means that it is unlikely to be a good representation of the change in average power generation. Essentially, the wave climate in the nearshore cannot be considered as simply a less energetic version of the offshore wave climate (if this were the case then there would be more justification for the use of the average omni-directional wave power as a proxy for average power generation). Alternative representations have been proposed to compensate for this potential distortion. Thus, there is the average directionally-resolved wave power, which is the wave power resolved to a particular direction and also the average exploitable wave power, which is where the directionally-resolved wave power is limited to a fixed multiple (typically four) of the average.

3.4   Measurement of Ocean Waves

3.4.1   Overview

Measurement of the ocean waves is clearly a very important part of understanding the wave energy resource. Although it is possible to estimate the expected waves at a particular location by analysing wave fields and modelling wave propagation, actual field measurements are required to both validate these estimates as well as provide information on the sea-state where an accurate model cannot currently be easily constructed. An example of the latter case would be measuring the impact of a wave energy converter on the down-wave conditions.

Table 3.1 contains a summary of the available systems and their characteristics.

Table 3.1 Wave measurement systems

System	Key characteristics
Surface-following buoy	Relatively expensive, accuracy well established, affected by currents, limited accuracy in steep waves, suitable for long-term deployment
Seabed pressure sensor	Relatively cheap, only suitable for shallow water, deployed in an array, can give directional characteristics, also measures variation in water depth
Acoustic current profiler	Relatively expensive, suitable for water depths up to 50 m, also measures marine currents, recovery required to extract data
Radar (land-based)	Deployed on land away for aggressive environment, typically requires calibration for each site, often limited to wave height measurement
Radar (satellite)	Large geographical coverage with low spatial and temporal resolution, typically limited to wave height measurement

The first three wave measuring instruments described above are deployed at the location of interest and use a time-series analysis to produce the wave spectrum. The duration used to define the wave spectrum depends on a number of competing factors. The first factor is that shorter sampling durations result in a larger uncertainty in the estimated wave spectrum and sea-state parameters. For example, the standard uncertainty in the significant wave height based on a 15-min sample in the North Atlantic is about 15 %, with the standard uncertainty reducing proportionally with the square root of the sample duration [18]. The second factor is that the sea-state is continually changing, with the waves growing or subsiding with the varying wind fields. Consequently, excessively long sample durations would potentially result in extreme sea-states being "smoothed out". The typical 15–30 min sample duration is generally considered to be a reasonable compromise between these two factors. The last of the factors is battery life and deployment time since powering the measuring instrument will limit the product of the sampling direction and number of sea-state records. For example, a wave-measuring buoy may be able to be used for a year deployment with 3 hourly records using 15 min sample durations, or a 2-month deployment reporting every hour using 30 min sample durations. The final choice of sample duration and frequency will depend on the purpose of the deployment, which should be carefully designed to ensure that the required information is obtained.

3.4.2 Surface-Following Buoy

Currently, the most common way of measuring ocean waves is with a **surface-following buoy** that is slackly moored to the seabed as shown in Fig. 3.17. In these buoys, the vertical motion is typically measured using an accelerometer (although GPS systems are now becoming more common), which can then be double integrated to provide a time-series of the water surface elevation [19]. The recorded surface elevation is then used to estimate the wave spectrum and from that the sea-state parameters. Wave measurement buoys may also contain instruments to measure the inclination of the buoy so that the direction of the waves can be inferred and used to produce a directional wave spectrum, with associated directional sea-state parameters. A major benefit of using wave measuring buoys is that they have been used for a long time and thus their accuracy is well established. Their limitations are also relatively well recognised in that strong currents and steep

Fig. 3.17 A
surface-following wave
measurement buoy

waves both reduce the accuracy of the measurements because the buoy does not exactly follow the water surface. In addition, they are relatively expensive and there is a relatively high risk of loss of the instrument.

3.4.3 Sea-Bed Pressure Sensor

A cheaper alternative to wave measuring buoys are **sea-bed pressure sensors** as shown in Fig. 3.18. These instruments measure the variation in pressure and from that infer the water surface elevation [20]. The attenuation of wave pressure with depth means that they are only suitable for relatively shallow water, with the high-frequency waves being attenuated the most and thus sea-bed pressure sensors are most suitable for measuring swell waves. The exact water depth limit for sea-bed pressure sensors depends on the signal-to-noise ratio of the instrument, but a depth limit of about 10–20 m is typical. In addition to the waves, a sea-bed pressure transducer can also be used to determine the tide level, although in this case care should be taken to ensure that changes to the atmospheric pressure are accounted for. Finally, sea-bed pressure sensors may be deployed in an array to provide information on the directional distribution of the waves.

Fig. 3.18 A seabed pressure transducer for wave measurement in shallow water

3.4.4 Acoustic Current Profiler

A more recently developed method of measuring waves is the use of **acoustic current profilers**, and in particular multi-beam acoustic current profilers as shown in Fig. 3.19. Acoustic current profilers measure the water velocity using the red/blue shift in acoustic pulses for the instrument [21]. The water velocities as determined from each beam are combined and processed to produce a time-series of the 3D wave-induced water velocities and from that the directional wave spectrum. Although these instruments can be deployed in any orientation, it is normal for wave measurement to deploy them on the seabed where they are less susceptible to damage and bio-fouling. In this case the instrument must be deployed in less than about 50 m of water since they become less capable of detecting the wave-induced

Fig. 3.19 An acoustic
current profiler

water velocities as they are further from the water surface. However, acoustic
current profilers are developing rapidly, and it is possible that future improvements
mean that they will be suitable for deployment in deeper water. An additional
advantage of acoustic current profilers is that they also provide information on the
local marine currents; however, a disadvantage that they share with sea-bed pressure transducers is that they typically store the data on board. This means that the
existence or quality of the wave resource data is not known until the instrument is
recovered, which could be a problem if the data is critical for project development.

3.4.5 Land-Based and Satellite Radar

In addition to local wave measurement, it is also possible to use remote sensors,
principally **radar**, to measure the waves, for example the 16-element HF radar array
set up on the dike at Petten (NL) as shown in Fig. 3.20. The main advantage of
using a remote sensor is that the instrument does not need to be deployed in an
aggressive environment, that the data is readily available and that the costs are
typically lower [22]. The radar provides information on the sea-state by analysing
the backscatter from the waves over an area. Coupled analysis from multiple
locations means that the directional wave spectrum can often be calculated, together
with sea-state parameters. Typically the radar system is calibrated using data from a

Fig. 3.20 A land-based radar
system for measurement of
waves

local wave measuring instrument [23]. The use of radar for wave measurement is a rapidly developing area, with a range of different systems being developed and continually improved.

A radar deployment of particular note is **in satellites**. Radar altimeters are available in a number of satellites that are circling the earth and can provide estimates of the Significant Wave Height along the track of the satellite [24]. Although the temporal resolution of the wave resource data may be very low (a satellite may only pass over a point every 10–35 days) it does provide wave resource data over a large geographical area. Although current satellite-based radar altimeter systems typically only provide information on the Significant Wave Height, as with other radar systems, the potential for satellite measurement of waves is increasing rapidly. Thus, current developments exist to use satellite data to produce estimates of the wave period as well as potentially other sea-state parameters.

Interestingly, the satellite data can be used to compare the output from local wave measuring instruments and the consistency of these assessed. Whilst the Canadian and UK wave buoy networks share a common calibration factor, the US buoy network has a slightly different calibration factor. This implies that there is some difference in the calibration, deployment and/or operation of the wave buoys between the US and Canada/UK that results in a difference in the estimated sea-state. This illustrates the challenges in wave measurement and that there are sources of uncertainty and error in all measurement systems. Consequently, there is no definitive "gold standard" for wave measurement, which is important to bear in mind for the calibration of wave propagation models that are discussed in the next section.

3.5 Modelling of Ocean Waves

3.5.1 Introduction

There are two main reasons for modelling ocean waves. The first reason is that until it is known where a wave energy converter may be deployed, it is unlikely that data from a wave measuring instrument for the desired location will exist. In this case a model is required to propagate the waves from where the wave resource data is known to the points of interest. The second reason is that knowledge of the average wave climate requires many years of data and it is not typically possible to deploy a wave measuring instrument for the time required to produce the required information, so a wave model is used to generate this long-term data.

Although other wave propagation models exist, such as Boussinesq [25] and mild-slope models [26], it is generally accepted that third generation spectral wave models are most suitable for modelling the propagation of waves from long-term wave measurement points or validated deep-water models to the potential locations of wave energy converters [27]. Examples of third generation spectral wave models include SWAN [28], TOMAWAC [29] and Mike21 SW [30], of which the first two are open source. Third generation spectral wave models solve the action balance equation shown in Eq. 3.23

$$\frac{\partial N}{\partial t} + \nabla_x \cdot [C_g + U]N + \frac{\partial C_\sigma N}{\partial \sigma} + \frac{\partial C_\theta N}{\partial \theta} = \frac{S_{tot}}{\sigma} \tag{3.23}$$

The action balance equation is essentially a conservation equation that states that the total action is conserved (left-hand side of the equation) except when there is an input of wave action (right-hand side of the equation) from sources/sinks of wave action. The wave action, which is the wave energy divided by the intrinsic wave frequency, is used because it is conserved in the presence of background currents whilst the wave energy is not. However, in the absence of background currents, the wave action balance equation reduces to the wave energy balance equation. The wave action is allowed to propagate between four dimensions, geographic space, frequency space, directional space and time space (dependent on the physical processes involved), but the change in the sum of these must equal the source/sink of wave action, which may generate, dissipate and/or re-distribute the wave action.

3.5.2 General Spectral Wave Models

In the absence of sources/sinks of wave action, then Eq. 3.23 simply defines the wave kinematics and includes the processes of both shoaling and refraction.

- Depth-induced shoaling is the change in wave height with water depth and, in the absence of marine currents, it occurs due to the change in group velocity with water depth. When a wave enters shallow water, the group velocity changes, initially increasing slightly and then decreasing as the water depth decreases. The consequence of this is that the wave height initially decreases and then increases due to the conservation of energy as the energy gets stretched out and subsequently bunched up (note, it is commonly assumed that wave height only increases in shallow water since the initial decrease in wave height is only small and often not considered; however, it does occur). Depth-induced shoaling normally refers to the case when the wave is propagating orthogonally to the depth contours and so the wave does not change direction. Assuming a time-invariant condition, depth-induced shoaling is all included in the second term of the wave energy balance equation.
- When the wave is not travelling orthogonally to the depth contours, then depth-induced **refraction** occurs. Depth-induced refraction is also related to the change in wave group velocity with water depth. Indeed, it is possible to consider depth-induced shoaling as simply a special case of depth-induced refraction where the direction of wave propagation is parallel to the water depth gradient. If this is not the case, then the direction of wave propagation changes as the group velocity changes. Thus, when the wave group velocity starts to

reduce in shallow water, the wave tends to rotate so that the wave crests become more parallel to the depth contours. The rate of change on direction is defined by Snell's law, which relates the change in direction of propagation with the change in wave group velocity. The consequence is that when the waves reach the shore, they are nearly parallel to the shoreline (they do not get to be completely parallel to the shoreline because non-linear processes cause them to break before they get to that point). Assuming a time-invariant condition, this process is modelled with the second and fourth terms of the wave energy balance equation.

- Where there are marine currents, then **current-induced shoaling** occurs when there is a change in the background current. The kinematics are more complex in this case because the change in current also causes a change in the relative (or intrinsic) frequency, which is the frequency of the waves as observed from a reference frame travelling at the same velocity as the current. In addition, there is an exchange of energy between the current and the waves, which is the reason Eq. 3.23 is not defined in terms of wave energy, but in terms of wave action, which is conserved in the presence of changing currents. The exact current-induced shoaling can be determined by solving Eq. 3.23 which shows that a following current tends to reduce the wave height as the energy is spread out, whilst an opposing current causes the wave height to increase as the energy bunches. If the opposing current is sufficiently strong, this can stop the waves propagating, and the energy increases until the waves start to break and thus lose energy. This can be seen commonly at the mouth of rivers where there is area of breaking waves as the current stops the waves from travelling upstream. For a time-invariant system, this is modelled with the second and third terms of the wave energy balance equation.

- As would be expected, marine currents can also be responsible for **current-induced refraction**. The explanation for this is similar to that of the depth-induced refraction, except that the change in velocity of the wave crest is due to spatial variation of the marine currents, rather than the change in water depth. In this final case, a time-invariant system the second, third and fourth terms of the wave energy balance equation are required to solve the wave kinematics.

- A major limitation of spectral wave models is that they assume a random **phase between the wave components**, although this is also the assumption that allows them to be computationally efficient for modelling wave transformations over large distances. Consequently, phase-dependent processes such as harbour resonance and diffraction cannot be modelled explicitly. However, whilst no approximation for processes such as harbour resonance exist for spectral wave models, it has been found that diffraction can be modelled using a phase-decoupled refraction-diffraction approximation [31]. This approximation is based on the mild-slope equation where the turning rate of the wave due to diffraction is dependent on the slope of the wave amplitude, and has been found to be a reasonable approximation in the majority of circumstances except locally close to the body causing the diffraction field.

3.5.3 Third Generation Spectral Wave Models

In third generation spectral wave models (for example SWAN, Mike21SW, TOMAWAC), there are typically six types of **wave energy source/sink**. These are wind input, quadruplet wave-wave interactions, white-capping, bottom friction, triad wave-wave interactions and depth-induced wave breaking, cf. Fig. 3.21.

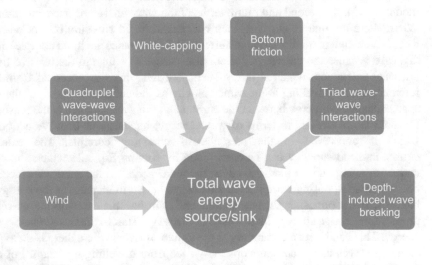

Fig. 3.21 Source terms used in a third-generation spectral wave model

These sources/sinks are, in general, represented by semi-empirical approximations. In many cases, a number of alternative representations can be used in the wave models; the choice being somewhat dependent on the preferences of the modeller. A brief description of each of these source/sinks of wave energy is provided; however, further details may be found in the literature [27].

- The wind input source term represents the energy that is transferred from the **wind into the waves**. Energy is primarily transferred through the propagation of pressure fluctuations that travel in the same direction as the waves. In the initial generation of waves, these pressure fluctuations cause small ripples on the surface of the water. Subsequently, the waves influence the air-flow so that there is a positive pressure on the windward side of the wave crest and a negative pressure on the leeward side. Thus, the net force on the wave and the wave velocity are in-phase. There is a transfer of energy from the wind to the wave and so the waves grow. Moreover, this net force increases with the size of the wave so that there is a positive feedback mechanism in the growth of the waves. However, whilst the general processes of wind to wave energy transfer may be understood, estimates of the wind input source term are based on a semi-empirical representation of the processes discussed above.

- The energy input from the wind to the waves occurs towards the high frequency side of the spectral peak, with the energy subsequently transferred to lower frequency waves (together with a small proportion of the energy being transferred to higher frequency waves) via **non-linear quadruplet wave-wave interactions**. The quadruplet wave-wave interactions are associated with multiple resonant couplings between sets of four wave components that cause energy transfer via non-linear interactions. Thus, quadruplet non-linear wave-wave interactions are responsible for the increase in the average wave period of the waves so that as the wind blows both the wave height and period increase. However, quadruplet wave-wave interactions also tend to stabilise the high frequency components of the spectrum so that in the absence of wave breaking, a fourth-order frequency tail (f^{-4}) is produced. Unfortunately, calculation of the quadruplet wave-wave interactions is computationally very expensive and so it is usual to use an approximation to calculate the strength of the quadruplet wave-wave interactions source term. Finally, it is worth noting that the explicit calculation of the quadruplet wave-wave interactions is the key distinguishing feature of third-generation spectral wave models compared to first and second order spectral wave models that either assume or parameterise the spectral shape.
- The final process that controls the spectral shape in deep water is **white-capping**. White-capping is wave breaking that occurs in deep water and may be expected to be associated with the steepness of the waves. That is, when the steepness of a wave becomes too large, the top of the wave becomes unstable and the wave breaks resulting in white-capping. However, as the wave steepens, the hydrodynamics become highly non-linear, and a complete theoretical understanding of white-capping has not yet been developed. Notwithstanding this lack of understanding, the strength of the white-capping source term is typically based on a model where the weight of the white-cap extracts energy from the waves in the opposite sense of the wind input source term, i.e. the net force on the wave due to the white-cap extracts energy from the waves. The effect of white-capping tends to be strongest on the high frequency wave components as these are typically steeper and result in a fifth-order frequency tail (f^{-5}) as seen in the JONSWAP and many other standard spectra.
- As the waves enter shallow water, they will begin to be affected by the seabed, with the most obvious (although not necessarily most significant) source term being **bottom friction**. The bottom friction source term represents the energy transfer from the waves to turbulence induced by shear stress from fluid flow over the bottom. A quadratic relationship between shear stress and fluid flow is typically assumed, as in Morison's Equation, so that the strength of the bottom friction source term is proportional to the square of the wave-induced velocity at the seabed. The coefficient of proportionality, or bottom-friction coefficient, is dependent on the characteristics of the seabed, with rocky, sandy and vegetated seabeds all having different bottom-friction coefficients.
- Another process that becomes more significant as the waves enter shallow water is the **triad wave-wave interaction**. In a similar way to quadruplet wave-wave interactions, triad wave-wave interactions are associated with multiple resonant

couplings between wave components, but in this case they are between sets of three wave components that cause energy transfer via non-linear interactions. When the waves are dispersive these interactions cannot be created, which is why they only occur in very shallow water. The effect of triad wave-wave interactions is to generate a second peak in the wave spectrum at twice the frequency of the original spectrum, which is bound to the main frequency peak in the sense that it travels with the same phase velocity. Unfortunately, currently third generation wave models are unable to correctly model these bound waves, and the triad wave-wave interaction source term is estimated based on the wave spectrum and water depth; however, these source terms have been found to be reasonable approximations in most cases.

- The final source term typically included in wave models is the **depth-induced wave breaking** (surf-breaking) source term. This source term typically assumes that a fixed proportion of the energy in any wave is lost when it breaks. Thus it is necessary to make an estimate on the proportion of waves that break at any particular water depth for any particular wave spectrum. This may be done by making some assumptions about the distribution of wave heights and the relative water depth in which these waves will break. Perhaps surprisingly, laboratory observations suggest that the spectral shape is not affected by wave breaking and so the energy removed due to wave breaking, and thus the strength of the depth-induced wave breaking source term, is typically assumed to be proportional to the wave energy spectrum, with the coefficient of proportionality dependent on the proportion of breaking waves.

3.5.4   Grid Definition

With the definition of the conservation equation for the action density, together with the source terms that add and remove energy, it is only necessary to propagate the action density from one point to another in order to define the wave resource. This may be done using a regular or irregular grid, using a number of different propagation schemes, each of which have their own particular set of advantages and disadvantages. The resolution of the grid will depend on the desired accuracy of the model, with higher resolution grids typically resulting in more accurate models, but at the expense of greater computational effort. However, the change in accuracy with grid resolution also depends on the relative magnitude of the components in the action density equation, so that where the action density changes little a low resolution grid may be used, but higher resolution grids are required where the changes in action density are greater. This naturally leads to the use of irregular (or nested regular) grids as shown in Fig. 3.22, which can have a higher resolution where required (although the areas requiring higher resolution grids and the required resolution may not be immediate obvious).

Fig. 3.22 An example grid
for a spectral wave model

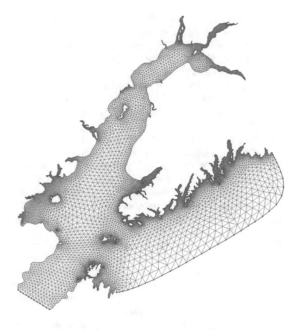

Combining the uncertainty in the grid design with the uncertainties in the model boundary data (e.g. the waves on the model boundary, the bathymetry, the variation in the seabed bottom friction coefficient, the choice of source term models), it can be seen why setting up a spectral wave model has been likened to a "black act". Unfortunately, there is no simple procedure that can be followed to guarantee the accuracy of the numerical wave model. A good match between the model prediction and measured data may provide some confidence in the model. However, this is strictly limited to the particular sea-state and the model may be less accurate when applied to different wind and wave conditions. Thus, it is generally recommended that any numerical model should be validated using a wide range of different conditions that include what may be expected over the whole year. Unfortunately, this data is not always available, and the wave resource model may be validated against only a limited data-set or even no data-set at all. In general, and in these cases in particular, it is clearly necessary to be aware that the wave resource is not fully validated and the WEC performance based on this data should be treated as such.

References

1. World Meteorological Organisation: Guide to wave analysis and forecasting (WMO 702), 2nd edn. WMO (1998)
2. Garrison, T.S.: Oceanography: An Invitation to Marine Science, 7th edn. Brooks/Cole, Belmont, USA (2010)

3. Barstow, S., et al.: WorldWaves wave energy resource assessments from the deep ocean to the coast. In: 8th European Wave and Tidal Energy Conference Uppsala, Sweden (2009)
4. SI Ocean: Resource Mapping—Work Package 2. DHI (2014)
5. Gunn, K., Stock-Williams, C.: Quantifying the global wave power resource. Renew. Energy **44**(0), 296–304 (2012)
6. Dean, R.G., Dalrymple, R.: Water Wave Mechanics for Engineers and Scientists. World Scientific Publishing Ltd., Singapore (1991)
7. Le Méhauté, B.: Introduction to Hydrodynamics and Water Waves. Springer, New York (1976)
8. Whittaker, T., Folley, M.: Nearshore oscillating wave surge converters and the development of oyster. Philos. Trans. R. Soc. A **370**, 345–364 (2012)
9. Folley, M., Whittaker, T.J.T., Henry, A.: The effect of water depth on the performance of a small surging wave energy converter. Ocean Eng. **34**(8–9), 1265–1274 (2007)
10. IEC/TS 62600-101: Wave energy resource assessment and characterisation, Marine Energy— Wave, tidal and other water current converters (2015)
11. Massel, S.R.: Ocean Surface Waves: Their Physics and Prediction, 2nd edn. World Scientific Publishing, Singapore (2013)
12. Pierson, W., Moskowitz, L.: A proposed spectral form for fully developed wind seas based on the similarity theory of S.A. Kitaigorodskii. J. Geophys. Res. **69**, 5181–5203 (1964)
13. Hasselmann, K., et al.: Measurement of wind-wave growth and swell decay during the joint North Sea wave project (JONSWAP). Dtsch. Hydrogr. (1973)
14. Battjes, J.A.: Surf similarity. In: 14th International Conference on Coastal Engineering. Copenhagen, Denmark (1974)
15. Komar, P.D.: Beach Processes and Sedimentation, 2nd edn. Prentice-hall (1998)
16. Folley, M., Whittaker, T.J.T.: Analysis of the nearshore wave energy resource. Renew. Energy **34**(7), 1709–1715 (2009)
17. Folley, M., Elsaesser, B., Whittaker, T.: Analysis of the wave energy resource at the European Marine Energy Centre. In: Coasts, Marine Structures and Breakwaters Conference. ICE, Edinburgh (2009)
18. Tucker, M.J., Pitt, E.G.: Waves in Ocean Engineering. Elsevier (2001)
19. Datawell: Datawell Waverider Reference Manual (2009)
20. Howell, G.L.: Shallow water directional wave gages using short baseline pressure arrays. Coast. Eng. **35**, 85–102 (1998)
21. Work, P.A.: Nearshore directional wave measurements by surface-following buoy and acoustic Doppler current profiler. Ocean Eng. **35**(8–9), 727–737 (2008)
22. Wyatt, L.R.: High frequency radar applicatoins in coastal monitoring, planning and engineering. Aust. J. Civil Eng. **12**(1), 1–15 (2014)
23. Siddons, L.A., Wyatt, L.R., Wolf, J.: Assimilation of HF radar data into the SWAN wave model. J. Mar. Syst. **77**(3), 312–324 (2009)
24. Mackay, E.B.L., Retzler, C.: Wave energy resource assessment using satellite altimeter data. In: 27th International Conference on Offshore Mechanics and Arctic Engineering. Estoril, Portugal (2008)
25. Bayram, A., Larson, M.: Wave transformation in the nearshore zone: comparison between a boussineq model and field data. Coast. Eng. **39**, 149–171 (1999)
26. Porter, D.: The mild-slope equations. J. Fluid Mech. **494**, 51–63 (2003)
27. Holthuijsen, L.H.: Waves in Oceanic and Coastal Waters. Cambridge University Press, Cambridge (2007)

28. SWAN: SWAN Technical Documentation. Delft University of Technology, Delft (2007)
29. EDF: TOMAWAC: Software for Sea State Modelling on Unstructured Grids Over Oceans and Coastal Seas. Release 6.0 (2010)
30. DHI: Mike21 Spectral Wave Module—Scientific Documentation. DHI (2008)
31. Holthuijsen, L.H., Herman, A., Booij, N.: Phase-decoupled refraction-diffraction for spectral wave models. Coast. Eng. **49**(4), 291 305 (2003)

Chapter 4
Techno-Economic Development of WECs

Arthur Pecher and Ronan Costello

4.1 Introduction

4.1.1 Continuous Evaluation of the WEC Potential

The development of a WEC, from having a good idea to demonstrating a commercially viable WEC, is an exciting but challenging journey. If the technology is right, it is expected to take about 15 years and a cost of two-digit million euros (in the best case) [1]. However, most of the technologies that are being developed will most likely never reach commercialisation, because they are not capable of producing market competitive electricity or they do not manage getting the required funding to proceed with development.

In order to avoid wasting large amounts of resources into the development of a technology, its potential of producing electricity at market price needs to be assessed continuously. Whenever the resulting LCoE calculation at the end of a development phase concludes that it is not sufficient for successful commercialisation, then there is little reason to proceed with its development. The chance that the further development of the technology will lower the LCoE is very small, while chances are rather high that unexpected cost will occur and, thereby, the final LCoE will be higher. If this unpleasant situation should occur, the fundamentals of the technology have to be readdressed, bringing the development of the technology back to the research phases [2].

A. Pecher (✉)
Department of Civil Engineering, Aalborg University, Thomas Manns Vej 23, 9220 Aalborg Ø, Denmark
e-mail: apecher@gmail.com

R. Costello
Wave Venture Ltd., Unit 6, Penstraze Business Centre, Truro, TR4 8PN, UK
e-mail: ronan@wave-venture.com

© The Author(s) 2017
A. Pecher and J.P. Kofoed (eds.), *Handbook of Ocean Wave Energy*,
Ocean Engineering & Oceanography 7, DOI 10.1007/978-3-319-39889-1_4

81

The calculation of the actual LCoE from a commercially operated WEC power plant relies on many assumptions and estimations. As long as a full-scale WEC has not been operated at the location of interest for a sufficient amount of time, there will be uncertainty in the power production and in the related costs. These uncertainties are larger, the further the technology is from commercialisation. So, throughout the development of a WEC, it is one of the main objectives to tackle these uncertainties. The further the development proceeds, the smaller these uncertainties will become and, thereby, a better estimation of the actual LCoE can be made. Various tools and calculations sheets are publicly available, which can facilitate the calculation of the LCoE [3–5].

4.1.2   Overview of the Techno-Economic Development

The technical performance level (TPL) and technical readiness level (TRL) scales, which are presented in Sect. 4.3.1, are especially used to rate the technical maturity (TRL) and economic potential (TPL) of a new technology and are very convenient as they facilitate the comparison between different developing technologies, even outside the wave energy sector. However, in practice a WEC is usually developed following a more specific set of development stages. These technical development phases of WECs are explained more in detail in Sect. 4.2 and can be coupled to the TRL scale [6].

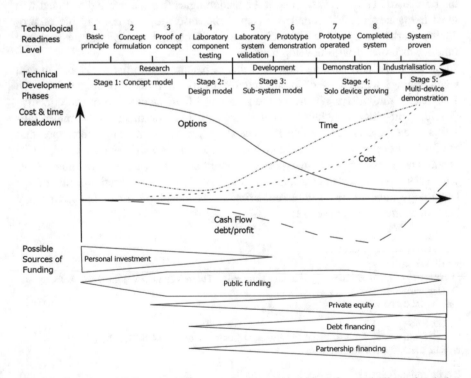

Fig. 4.1 Overview of different techno-economic parameters and how they typically evolve during the technical development of a WEC [1, 7–9]

Figure 4.1 presents an overview of how the time, cost and design parameters evolve with the technical development phases and, thereby, also the TRL. The main trends are that the required time and expenses increase significantly with increasing TRL while the amount of design variables that can be changed decrease significantly. The sources of funding usually tend to change significantly throughout the development process as well.

As no technology has yet been successfully commercialised, the current "best practice" is still based on experiences from other sectors and on assumptions from experts in the wave energy sector. Thereby, the details in the Fig. 4.1 remain approximate.

Note that the development strategy, which can be strongly influenced by the financing body, has a significant influence on the cost and time break-down over the different development stages. Some might favour spending additional time at the research level where all options are still open and time and cost are relatively small, only to proceed when a sufficient LCoE level (TPL) has been reached. Other financing bodies might favour a faster (but more risky) development process in which the TRL prevails. Some different possible development strategies are discussed in Sect. 4.4.4.

4.2   The WEC Development Stages

The technical development of a WEC can generally be divided into 5 main development stages [10]. Each stage is characterised by very specific goals and objectives which make it possible to progress systematically. As the development of WECs is very time-consuming and capital intensive, it is a challenge to keep these to a minimum. However, proceeding too quickly in a phase or even missing a phase can and will most likely have significant negative repercussions on the further development of the WEC. Note that modifications to the concept or design of the WEC should be done as early as possible through the development as this will become only more difficult, costly and time-consuming if they are done at a later stage.

The following Table 4.1 presents the different development stages that characterise a typical development path of WECs, from idea to commercialisation. It includes the main characteristics of each of them. Note that:

- Each development stage requires specific WEC model(s)/prototype(s) that will be subjected to specific challenges and objectives.
- From development phase 3, no significant changes to the overall WEC configuration are supposed to be made, thereby proceeding from the research to the development.
- The power production outcomes from laboratory tests should rely on tests in representative wave conditions for locations of interest.
- At the end of each phase, the progress and LCoE have to be evaluated. Based on this, a decision is made on whether the development can be taken to the next phase or if it is even worth continuing the development of the WEC.

Table 4.1 Detailed overview of the five development stages for wave energy converters [7, 9, 11]

	Stage 1: Concept model	Stage 2: Design model	Stage 3: Functional model	Stage 4: WEC prototype	Stage 5: Array demonstration
Illustration					
Scale	1:20–1:100	1:10–1:50	1:3–1:10	1:1–1:3	1:1
Location	Laboratory	Laboratory	Laboratory/benign site	Open seas	Open seas
Model/Prototype characteristics	– Idealized setup – Load-adaptable PTO – Adaptable design variables	– Final design – Representative characteristics – Simulated PTO	– Full fabrication – True PTO and – Electrical generator	First fully operational device	Autonomous and operational WEC power plant
Waves	Representative power production and extreme sea states	Representative power production and extreme sea states	Pilot site waves	Operational and extreme sea states	Operational and extreme sea states
Experimental test objectives	Main – Concept validation and optimisation – Power performance estimation – Assessing the impact of design variables and environmental parameters Possibly also – PTO and mooring char. – Loads estimation – Movement estimation (RAO's)	Main – Power performance estimation – Mooring and structural loads – Sea keeping – PTO conditions – Assessing the impact of design and environmental variables Possibly also – Detailed numerical calc. – Feasibility study	Providing experimental data and experience on – Power performance – Wave-to-wire model, including control strategy – Mooring and structural loads – Survival and sea keeping – Marine environment	Real cost and power production data for projection for device sales – CapEx – OpEx – Energy production And also – Wave-to-wire model – Structural and mooring forces – Lifecycle assessment	Real cost and power production data for projection for WEC array sales: – Array CapEx – Array OpEx – Array Energy production And also for WEC array – Wave-to-wire model – Structural and mooring forces – Lifecycle assessment

Techno-economic development of WECs

4.3   Techno-Economic Development Evaluation

4.3.1   The Technology Readiness and Performance Level

Recent work to provide ways of measuring the progress and the value of technology R&D processes has focused on adapting the TRL to specific wave energy terms and the introduction of a new TPL scale.

ESBI and Vattenfall [12] have prepared the wave energy TRL scale focusing on functional readiness and lifecycle readiness. While, Weber [1, 13] has prepared the TPL scale focusing on an all-round performance assessment with heavy emphasis on innovation and assessing economic viability. Additional wave farm TRL scales have been published [14] and a complimentary scale of Commercial Readiness Levels (CRL) has been defined to extend beyond the R&D phase [15].

Functional readiness means the readiness to convert ocean wave energy and export it to grid in addition to other related and essential functions such as station keeping and remote monitoring. The TRL scale gives indications of how these should be demonstrated at different TRL levels. Lifecycle readiness means readiness in non-functional areas that are important to utilities; these include operational readiness, supply chain readiness, risk reduction and also cost estimation and reduction. Inherent to the TRL scale is a focus on certification and a related expectation for the end user to be required to insure against certain risks (Table 4.2).

Table 4.2 The technological readiness levels (TRL)

TRL	Functional readiness	Lifecycle readiness
9	Operational performance and reliability demonstrated for an array of WECs	Fully de-risked business plan for utility scale deployment of arrays
8	Actual full-scale WEC completed and qualified through test and demonstration. (1:1 Froude)	Actual marine operations completed and qualified through test and demonstration
7	WEC prototype demonstration in an operational environment. (>1:2 Froude)	Ocean operational readiness: management of ocean scale risks, marine operations, etc
6	WEC prototype demonstration in a relevant environment. (>1:4 Froude)	Customer interaction: consider customer requirements to inform design. Inform customer of likely project site constraints
5	WEC component and/or basic WEC subsystem validation in a relevant environment. (>1:15 Froude)	Supply-chain mobilisation: Procurement of subsystem design, installation feasibility studies, cost estimations, etc
4	WEC component and/or basic WEC subsystem validation in a laboratory environment. (>1:25 Froude)	Preliminary lifecycle design: targets for manufacturable, deployable, operable and maintainable technology
3	Analytical and experimental critical function and/or characteristic proof-of concept	Initial capital cost and power production estimates/targets established
2	WEC concept formulated	Market and purpose of technology identified
1	Basic principles observed and reported	Potential uses of technology identified

The TPL scale is focused on performance as a combination of social, environmental and legal acceptability, power absorption and conversion, system availability, capital expenditure (CapEx) and operational expenditure (OpEx). Inherent to the TPL scale is a focus on Cost of Energy (CoE) and on improving this through innovation at low TRL. A further focus of the TPL is on formulation and automation of the performance assessments. An important component of the performance assessment is techno-economic simulation and optimisation; this ideally combines simulation of the physical processes in wave energy absorption with operational simulation, financial assessment and numerical optimisation techniques [16–18] (Table 4.3).

Table 3.4 The technological performance levels (TPL)

TPL	Category	Performance
9	High: Technology is economically viable and competitive as a renewable energy form	Competitive with other energy sources without special support mechanism
8		Competitive with other energy sources given sustainable support mechanism
7		Competitive with other renewable energy sources given favourable support mechanism
6	Medium: Technology features some characteristics for potential economic viability under distinctive market and operational conditions. Technological and/or conceptual improvements required	Majority of key performance characteristics and cost drivers satisfy potential economic viability under distinctive and favourable market and operational conditions
5		In order to achieve economic viability under distinctive and favourable market and operational conditions, some key technology implementation improvements are required
4		In order to achieve economic viability under distinctive and favourable market and operational conditions, some key technology implementation and fundamental conceptual improvements are required
3	Low: Technology is not economically viable	Minority of key performance characteristics and cost drivers do not satisfy potential economic viability
2		Some of key performance characteristics and cost drivers do not satisfy potential economic viability
1		Majority of key performance characteristics and cost drivers do not satisfy and present a barrier to potential economic viability

4.3.2 The WEC Development Stages and the TRL Scale

The five technical development stages (see Sect. 4.2) are specific to the wave energy sector while the TRL scale, which rates the technical maturity (see Sect. 4.3.1), is widely used in other industries.

Although these two systems are in some aspects very different, they can still be combined and compared as they both follow the development of a new product. This is presented in the following Table 4.4.

Table 4.4 The TRL and WEC development stages [12, 19]

WEC development phase	TRL	Short description	Model/prototype	Required funding	System fundamentals flexibility
1	1	Basic principles observed and reported	Scaled models or subsystems in relevant environment	+	+++++
	2	Technology concept and/or application formulated			
	3	Analytical and experimental critical function and/or characteristic proof-of-concept			
2	4	Technology component and/or basic technology subsystem validation in a laboratory environment		+	++++
3	5	Technology component and/or basic technology subsystem validation in a relevant environment			
	6	Technology system model or prototype demonstration in a relevant environment	Scaled prototype to full-scale WEC in sea trials	++	++
4	7	Technology system prototype demonstration in an operational environment		++++	+
4–5	8	Actual technology system completed and qualified through test and demonstration			
5	9	Operational performance and reliability of an array demonstrated	WEC array	+++++	+

4.3.3 The TRL-TPL R&D Matrix

As mentioned before, The TPL scale (from 1 to 9) presents the economic potential of a WEC while the TRL scale (from 1 to 9) presents the technical maturity level of a technology. These two evaluation scales can be combined in the TRL-TPL matrix, also called the Weber R&D matrix. This TRL-TPL matrix allows the status of a wave energy technology R&D programme to be represented as a point on the TRL-TPL plane and the history of the R&D progress up to that point as well as projections of future progress to be charted as lines.

A TRL-TPL matrix is presented in Fig. 4.2, in which the horizontal axis of the diagram is the **TRL** and the vertical axis is the **TPL**.

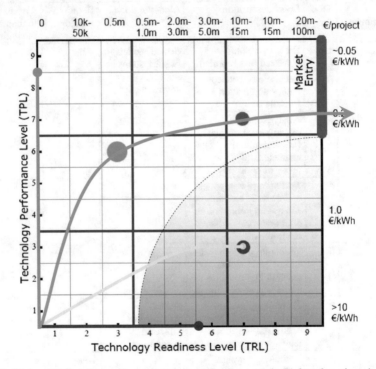

Fig. 4.2 Weber R&D Matrix. *Top edge* gives indicative spend. *Right edge* gives indicative performance levels. All R&D starts at *bottom left*. *Purple bar* is "market entry" the R&D goal. *Purple dot* is minimum viable product. *Green line* is an effective performance-before-readiness R&D trajectory. *Shaded area* is a "graveyard" for R&D programmes with low TPL. Adapted from [12] with permission

The right edge of the matrix is marked with indicative LCoE, which represents the TPL. Higher TPL levels are associated with more competitive cost of energy.

The top edge of the matrix is marked with indicative R&D spend or "burn rate". Higher TRL levels are associated with higher capital "burn rates" as the R&D expenditures and the project risks also increase dramatically with the TRL.

All technology developments enter the process at the left of the diagram and, if all goes well, proceed along a rightward and upward trajectory towards market entry. Successful market entry requires a fully developed WEC (TRL 9) that is commercially viable, meaning a TPL between 7 and 9 (with or without financial support).

The grey area represents the "graveyard". This area indicates the TRL-TPL combinations at which further developments should probably be ceased as it is very unlikely that from that point on the product will ever become economically viable. If the developer would, however, decide to proceed with the development, significant changes will have to be made to the basics of the concept, thereby returning to an earlier TRL in the hope to raise the TPL (see Sect. 4.4).

During the technical development of the WEC—in the form of experimental tests, numerical models and analysis—design decisions are made concerning the fundamentals of the concept. These WEC fundamentals are numerous and very flexible at an early stage as everything is still open for discussion while they are being addressed and, thereby, being fixed together with the development. Thereby, it is of great importance not to fix fundamental parameters of the WEC as long as the TPL is not at least 7 or above. Figure 4.3 presents the different domains of the TRL-TPL matrix.

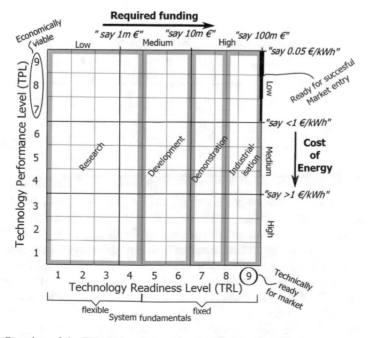

Fig. 4.3 Overview of the TPL-TRL matrix with related information [1]

While system fundamentals are flexible (left half of the diagram), the primary R&D goal should be to increase TPL with an emphasis on analysis, innovation and assessment of many alternatives and where this is facilitated by low cost and low risk activities. After a concept with sufficiently high TPL has been identified, the system fundamentals should be fixed and the R&D should progress to the right-hand side of the Weber diagram. In the right half of the diagram, the primary R&D goal is to increase TRL; in this domain the emphasis is on demonstration and risk reduction. In the right-hand domain, innovation must be much more cautiously managed to reduce risk in large projects and must be limited to improving sub-systems. Ideas for entire system improvements must be tested at lower TRL and treated as new projects.

4.3.4   Uncertainty Related to the TRL-TPL Matrix

As stipulated before, the LCoE for a commercially-operated power plant, based on a particular WEC, should be estimated at the end of each development stage. During the development of a WEC, the numerous assumptions and unknowns related to the cost and power production are addressed systematically. Thereby, the uncertainty related to the LCoE, which is a function of the cost and power production of a WEC, gets gradually reduced with the development phases. Table 4.5 presents EPRI attempt to quantify the level of uncertainty related to the estimated cost based on the technology's design maturity.

Table 4.5 Cost estimate rating table showing cost uncertainty as a percentage [20]

Design maturity \calculation detail	1. Conceptual (idea or lab)	2. Pilot	3. Demonstration	4. Commercial	5. Mature
A. Actual	–	–	–	–	0
B. Detailed	–	–	–15 to +20	–10 to +10	–5 to +5
C. Preliminary	–30 to +50	–25 to +30	–20 to +20	–15 to +15	–10 to +10
D. Simplified	–30 to +80	–30 to +30	–25 to +30	–20 to +20	–15 to +15
E. Goal	–30 to +200	–30 to +100	–30 to +80	–30 to +70	–

The values in the Table 4.5 are unlikely to be generally applicable. However, they give a probable indication of the uncertainty linked to the estimated cost of a WEC project. The overall uncertainty related to a WEC project will even be greater as there is also a fair level of uncertainty linked to the power production, which depends on the environmental conditions and availability of the WECs.

Figure 4.4 gives an example of possible LCoE estimations that have been re-evaluated all along the technical development of a WEC. The optimistic and pessimistic LCoE estimations illustrate the uncertainty related to the mean LCoE estimation.

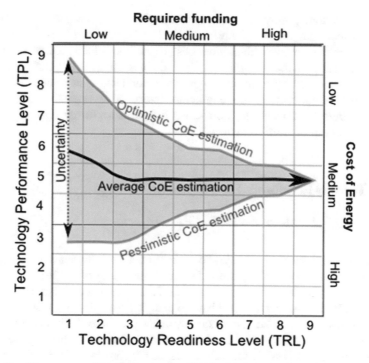

Fig. 4.4 Possible progress of the CoE estimation (TPL) with the technology development (TRL), together with an illustration of the potentially related uncertainty

The average LCoE estimation is the average between the optimistic and pessimistic LCoE estimation. Besides the uncertainty on the estimation, a different result might be obtained depending on who makes the calculation (e.g. developer or independent third-party). It is unfortunately difficult to estimate the fully correct LCoE in any case before a commercially operated power plant based on a particular WEC is built and operated over its full lifetime. So, it is of great importance that the LCoE estimation is transparent where (possible) assumptions are disclosed.

4.3.5   Valuation of R&D Companies

A further use of the Weber R&D matrix is as a guide for assessing the technology companies which are half-way through an R&D programme. For example, consider the different R&D programmes represented by a TRL = 7 and TPL = 3 and another case with a TRL = 3 and TPL = 6 (see dots in Fig. 4.2); imagine that the companies conducting these R&D programmes are raising equity. Which is more investible?

Conventional wisdom might argue that the higher TRL programme is closer to market readiness and, therefore, that the additional investment needed to bring the R&D to completion is less than in the case of the lower TRL programme. If an assessment is done purely on the basis of the TRL, then the dark blue dot would appear to represent the more advanced R&D programme. However, as already established in the previous sub-section, this programme is likely to stall or at least to have to go back to the drawing board: it finds itself in the "R&D graveyard". Conversely, the light blue dot, although at a lower TRL, is at a much higher TPL and crucially is much closer to the green trajectory. A valid relative measure of an R&D programme is, therefore, how close it is to a trajectory that will result in successful (affordable) market entry.

4.4 Techno-Economic Development Strategies

4.4.1 R&D Strategy as TRL-TPL Trajectories

An R&D manager has to choose the allocation of resources between achieving readiness before performance or performance before readiness. A readiness-before-performance trajectory would involve progressing along the TRL scale first and then along the TPL scale, so an R&D programme would have to complete multiple design iterations at high TRL and high cost and would be consequently unlikely to succeed. The horizontal red line represents an extreme version of this trajectory while the yellow line represents a less extreme version. It is possible for such development efforts to achieve a midlevel TRL, using a combination of private funding and public grant support. However, at higher TRLs the increased cost of R&D attracts greater levels of due diligence and such an effort would stall due to low TPL estimates in due diligence. The lower right area of the matrix is a "graveyard" for R&D programmes that rush through the early TRL stages too quickly and do not focus on achieving a high TPL while still at low TRL and low cost of design iteration. The orange and green lines are performance-before-readiness trajectories. The vertical orange line represents a trajectory that corresponds to a pure thought experiment; a WEC concept that never leaves the log book or imagination of the inventor. In principle, it is possible for this trajectory to reach high TPL, but with very high uncertainty in the TPL since no physical testing is done. This trajectory is not practical because it remains at very low TRL for too long; testing at TRL 2 & TRL 3 is needed to reduce uncertainty in the assessments in the early stages of the R&D effort. The green line represents a more practical version of the performance-before-readiness trajectory.

A trap to be avoided is attempting readiness before performance strategy in the belief that performance can be increased after market entry in line with anticipated learning rates. This strategy can be successful only in cases where (i) initial investment is sufficient to reach market entry and (ii) the product is viable so that customers buy multiple generations, and learning rates can come into consideration. In wave energy, neither of these conditions are likely to occur. A readiness before performance strategy is almost certain to fail in reaching market entry while a performance before readiness strategy will deliver a viable product more cheaply than any other strategy.

Weber [13] argues that the rapid increase in TPL is made possible by structured innovation techniques such as TRIZ [21] and techno-economic optimisation [22] applied at low TRL. A key requirement to success in this stage is flexibility in concept definition. The performance-before-readiness strategy facilitates this because radical changes to system fundamentals—e.g. from a point absorber to a terminator or from a submerged to a surface piercing device—are affordable and manageable at lower TRL. Conversely, at high TRL such changes would be prohibitively expensive, risky and would actually violate several guidelines for WEC development [23–25]. A consequence of the focus on flexibility and concept level innovation is that it may be necessary to test several or even many concepts to TRL 2 or TRL 3 in order to choose between alternatives for further development. A challenge in implementing the performance-before-readiness development trajectory is related to dealing with uncertainties in understanding the characteristics of the mature system before that system is actually available. This translates into a requirement for sophisticated techno-economic assessment and optimisation software for judicious use of experimental testing and, most importantly, for a structured approach to the innovation.

4.4.2   Extreme Cases of Techno-Economic Development Strategy

The techno-economic development strategy for a WEC might differ with respect to the importance of TPL or TRLs. Some extreme cases of techno-economic development strategies could favour one of them radically above the other [1], meaning that the WEC developer would prioritise:

- A rapid technology development of the WEC without addressing the technology performance. Here, the WEC developer will try to minimise the duration between the (initial) development phases. This strategy will be referred to as "Readiness before Performance".
- The performance of the WEC needing to be optimised before proceeding to the next development phase. Here, no progress in terms of development stage is made as long as the highest TPL, where no subsidies are required, is proven to be within reach. This will be referred to as "Performance before Readiness".

The adopted development strategy is usually a result of the different opinions and agendas of the different stakeholders behind the WEC, e.g. the inventor, (public and/or private) funding body etc., which might favour one strategy over the other. Table 4.6 presents an overview of the particularities of these two strategies.

Table 4.6 Overview of two extreme techno-economic development strategies

Readiness before performance	Performance before readiness
Characteristics	
– The WEC developer favours a quick development of the WEC, to limit the time spent at each TRL – The TPL is assumed to be sufficient, based on optimistic CoE estimations or on secondary importance – It is believed that the WEC fundamentals can be improved at a later stage, which is in practice very difficult, costly and time-consuming, and prior experience can become obsolete	– The WEC developer favours a thorough technical development of the WEC – At each TRL, the TPL is enhanced to optimal level, which is very time consuming and work intensive. However, when/if TRL 9 is reached the WEC is directly ready for successful market entry – The extensive work at each TRL also reduces the related uncertainty as all aspects have been carefully investigated
Possible argumentation	
– Being satisfied with its initial TPL, arguing it does not require any further development – Trying to rapidly become an important player in the sector – Trying to gain time (at early TRLs) believing it will also limit the related financial means – It is easier to attract interest (and funding) when the technical development is fast and the models/prototypes are larger	– Believing that the TPL is the most important, as there is no point to further develop a WEC that will not be competitive with other energy sources, without special support mechanism – That from the moment the highest TPL is reached, the rest (interest from investors etc.) will follow
TRL_TPL matrix illustration	

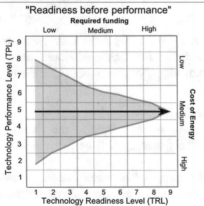

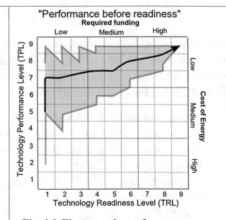

Fig. 4.5 Illustrates the readiness before performance development strategy

Fig. 4.6 Illustrates the performance before readiness development strategy

(continued)

Table 4.6 (continued)

Readiness before performance	Performance before readiness
(Probable) development pathway	
– The (first) development phases are passed rapidly, a lot of physical progress can be shown – At some point, it becomes relatively difficult to attract more funding, as the TPL does not justify the further development – Significant technical modifications are required to proceed with the development and in order to argue that the technology can become economically-viable – Here for, earlier development phases need to be repeated, requiring new designs, models and prototypes – This will be very expensive, time-consuming and making previous experience possibly obsolete. Moreover, the new design might still not lead to a commercially viable technology if the new fundamentals are not right	– The first development phases take a lot of time and effort as every aspect of the technology is carefully investigated – It might be difficult to get public attention, as progress is slow, models are small (at least during the first development stages) and the system might seem more complicated, as many more details have been investigated – It might be difficult to bridge the gap to sea trials; however, the value of the technology should become very clear the further it gets with the technical development – Once the technology can be demonstrated offshore in a reasonable size, most of the uncertainties should fade away and the commercial and technical potential should be clear – In case the technology, during its development, shows that it is not capable to reach TPL 9, then the technical development will be stopped and unnecessary time and cost will be avoided

In both cases, it can take a very long time to arrive at the required TPL and TRL to reach a successful market entry. For the "Readiness before Performance", the whole development will need to be repeated with updated WEC fundamentals while the "Performance before Readiness" will require substantial amounts of funding if the development duration becomes really long. The next sub-chapter will present a middle road.

4.4.3 Efficient Techno-Economic Development

First of all, the fundamentals of the WEC need to allow the technology to become commercially viable. This is of major importance and needs to be obvious and presentable at the end of each development phase. This might be a bit more difficult at early development phases as uncertainties are larger, but it should be well documented before sea trials take place. Therefore, all important aspects of the WEC technology, such as mooring, structural design, power production, survival mechanism, PTO design and others need to be assessed carefully in representative wave conditions before the WEC technology goes to sea trials.

Assuming that the fundamentals of the WEC in development are capable of bringing the WEC to a successful market entry, then the development trajectory should be optimised in order to limit the required amount of funding and the overall time to market. As changes to the WEC fundamentals are still flexible, relatively cheap and fast to change at early development stages, this should be the first priority. A lot of effort at relatively low cost can be dedicated in the beginning, e.g. optimising the Wave-to-Wire (W2W) performance and minimising the structural requirement, which can lead to substantial LCoE improvements. This will, in practice, mean various experimental test campaigns, using various models of the full system and of sub-systems separately so that the influence of a large range of physical and environmental parameters can be assessed. This will lead to an opti-mised design and an extensive knowledge of the loads and design characteristics of all essential parts of the device. The parallel development of a W2W numerical model can be highly valuable if it can be sufficiently accurate.

Figure 4.7 illustrates this efficient performance-before-readiness techno-economic development strategy.

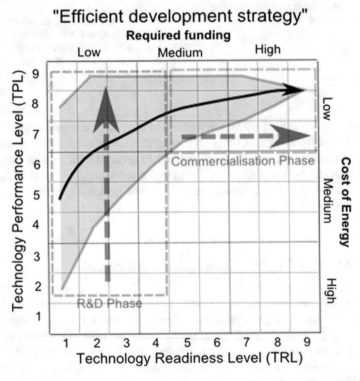

Fig. 4.7 Illustration of a possible successful and efficient techno-economic development of a WEC

Once the early development phases (mainly research) have maximised the TPL, the focus should be put on reducing the time to market in order to secure WEC sales income rather than further relying on external funding. This will be the start of the development process, which aims at demonstrating the WEC operating in a real sea environment. The first prototype will be of a reduced scale and operating in a benign site where the WEC can operate in reasonable wave conditions. The aim will be to have it be fully-functional and operating as an autonomous power plant unit. It should, however, also present storm condition data so that the storm configuration of the WEC can be assessed and extreme loads on the structure be measured. The last development stage will then present a full-scale WEC that is able to operate fully autonomously and that is ready for successful market entry. There will always be room for improvements, and they will have to be addressed in parallel with the commercial activities of the WEC company as any technology-based company do.

4.5   Conclusion

The successful development process of WECs demand large amounts of time and means. The optimal development trajectory manages to keep these expenses to a minimum while delivering an economically viable product at the end of its development. As related expenses (time and money) increase exponentially with the development stages (TRLs) while flexible parameters decrease rapidly, it is of the highest importance to optimise the WEC principles at an early stage (TRL 1-4) up to the level where the economic potential of the WEC is ensured (TPL > 7).

If, when passing TRL 4 (working principles of the WEC are fixed and the new outlook is demonstration), the TPL is not greater than 7 (at least economic viable with incentives), then the subsequent expenses will be wasted and could possibly harm the credibility and/or image of the technology developer or even the sector. In general, during each TRL of the development, the potential of the WEC of being capable of achieving successful market entry (TPL > 7) has to be assessed, taking the uncertainty with this estimation into account. If this turns out to be negative or indicates doubts relative to its potential, then the progress in terms of TRL should be stopped and it might even be required to go some development steps back. This will be the only option, as significant modifications to the WEC fundamentals are only possible at early TRLs.

When looking at the WECs currently being developed internationally, the working principles of WECs are still very broad (see Chap. 2) while only a very small fraction of these are expected to be able to reach the satisfactory TPL for successful market entry. These WECs in development have often rushed too quickly into the TRLs as they have produced too optimistic estimations of their TPLs (or they did not take the importance of the TPLs seriously).

4.6 Overview of Some of the Leading WECs

Table 4.7 presents an overview of some of the leading WEC technologies. These are indicative numbers shared by the corresponding companies at some point in the past. More WEC technologies could have been added, such as Wavestar, AW energy's Waveroller, AWS, Fred Olsen, Weptos, Seabased and possibly many others.

Table 4.7 Overview of some key figures of the development of WEC [26–35]

Company/model	OPD Pelamis	Aquamarine Oyster	OPT PowerBuoy	Oceanlinx	Carnegie CETO
Development start (year)	1998	2001	1994	1997	1999
Duration to first approx. full-scale prototype	6 years (2004)	7 years (2008)	11 years (2005)	9 years (2006)	11 years (2011)
Total Funding	Approx. 70 m £ Till stage 5 (2011)	Approx. 34 m £ Till stage 4 (2013)	52.8 m $ Till stage 4 (2011)	86 m AUD Till stage 4 (2011)	70 m AUD Till stage 4 (2014)
Amount in €	Approx. 64 m €	Approx. 41 m €	Approx. 38 m €	Approx. 58 m €	Approx. 47 m €
Estimated TRL (2014)	8	7	6–7	7	7–8
Comments on the development path	Redesign of the WEC at development phase 5	Redesign of the WEC at development phase 4	Redesign of the WEC at development phase 4	Redesign of the WEC at development phase 4	Redesign of the WEC at development phase 4
Reference	[26, 27]	[28]	[29–31]	[32, 33]	[34, 35]

It would have been great to be able to extend Table 4.7 with a TPL rating for each technology. However, these values and there underlying calculations and assumptions are rarely publically shared by developers.

References

1. Weber, J.: WEC technology readiness and performance matrix—finding the best research technology development trajectory. In: 4th International Conference on Ocean Energy, Dublin (2012)
2. Kurniawan, A.: Modelling and geometry optimisation of wave energy converters (2013)
3. Chozas, J.F., Kofoed, J.P., Helstrup, N.E.: The COE Calculation Tool for Wave Energy Converters (Version 1. 6, April 2014) (2014)
4. RETScreen International: Clean Energy project analysis software tool. http://www.retscreen.net/ang/home.php
5. Ocean Energy Systems (OES): International Levelised Cost Of Energy for Ocean Energy Technologies (2015)
6. Mankins, J.C.: Technology Readiness Levels: A White Paper (1995)
7. Holmes, B.: OCEAN ENERGY: Development and Evaluation Protocol—Part 1 : Wave Power, HMRC—Mar. Inst., pp. 1–25 (2003)

8. IEE: State of the art analysis—a cautiously optimistic review of the technical status of wave energy technology. Rep. Waveplam, Intell. Energy Eur. Brussels (2009)
9. Pecher, A.: Performance Evaluation of Wave Energy Converters. Aalborg University (2012)
10. Johnstone, C.M., Mccombes, T., Bahaj, A.S., Myers, L., Holmes, B., Kofoed, J.P., Bittencourt, C.: EquiMar : Development of Best Practices for the Engineering Performance Appraisal of Wave and Tidal Energy Converters
11. Kofoed, J.P., Frigaard, P.: Development of wave energy devices: The Danish Case. J. Ocean Technol. 4(2), 83–96 (2009)
12. Fitzgerald, J., Bolund, B.: Technology Readiness for Wave Energy Projects; ESB and Vattenfall classification system. In: 4th International Conference on Ocean Energy, pp. 1–8 (2012)
13. Weber, J., Costello, R., Mouwen, F., Ringwood, J., Thomas, G.: Techno-economic WEC system optimisation—methodology applied to Wavebob system definition, pp. 1–5
14. Neary, V., Lawson, M., Previsic, M., Copping, A., Hallett, K., LaBonte, A., Rieks, J., Murray, D.: Methodology for design and economic analysis of marine energy conversion (MEC) technologies. In: Marine Energy Technology Symposium (2014)
15. Australian Renewable Energy Agency: Commercial Readiness Index for Renewable Energy Sectors. (2014)
16. Weber, J., Costello, R., Ringwood, J.: WEC technology performance levels (TPLs)—metric for successful development of economic WEC technology. In: 4th International Conference on Ocean Energy (2012)
17. Teillant, B., Costello, R., Weber, J., Ringwood, J.: Productivity and economic assessment of wave energy projects through operational simulations. Renew. Energy 48, 220–230 (2012)
18. Padeletti, D., Costello, R., Ringwood, J.V.: A multi-body algorithm for wave energy converters employing nonlinear joint representation. In: Proceedings of the ASME 2014 33rd International Conference on Ocean, Offshore and Arctic Engineering, pp. 1–6 (2014)
19. Westwave: Appendix 2 Technology Readiness Levels for Supply Chain Study for WestWave (2011)
20. Previsic, M., Bedard, R.: Yakutat Conceptual Design, Performance, Cost and Economic Wave Power Feasibility Study (2009)
21. Gadd, K., Goddard, C.: TRIZ for Engineers: Enabling Inventive Problem Solving (2011)
22. Costello, R., Teillant, B., Weber, J., Ringwood, J.V.: Techno-economic optimisation for wave energy converters. In: 4th International Conference on Ocean Energy, p. not paginated (2012)
23. Snowberg, D., Weber, J.: Marine and Hydrokinetic Technology Development Risk Management Framework," (September) (2015)
24. Det Norske Veritas: Guidelines on design and operation of wave energy converters (2005)
25. Davies, P., EMEC: Guidelines for Design Basis of Marine Energy Conversion Systems (2009)
26. Carcas, M.: Pelamis—current status and prospects. Presentation at 4th international seminar on ocean energy at EVE, Bilbao, Spain (2011)
27. Yemm, R.W., Henderson, R.M., Taylor, C.A.E.: The OPD pelamis WEC : current status and onward programme. Int. J. Ambient Energy 24(1) (2003)
28. Murray, M.: Making marine renewable energy mainstream. (2013)
29. Posner, B.M.: PT—company presentation (2011). www.oceanpowertechnologies.com
30. Stiven, T.: OPT. Presentation at 4th international seminar on ocean energy at EVE, Bilbao, Spain (2011)

31. Nock, H., Montagna, D.: Commercialization of wave power technology. In: 11th Annual Congressional Renewable Energy and Energy Efficiency EXPO, pp. 1–6 (2008)
32. Oceanlinx: Wave Energy at its best. Presentation at Cleantuesday, Paris, France (2011)
33. Baghaei, A.: The Technology and the Maui Wave Energy Project (2010)
34. Ottaviano, M.: Carnegie Roadshow Presentation. pp. 1–24 (2014)
35. Carnegie Wave: CETO Wave Technology (2008)

Chapter 5
Economics of WECs

Ronan Costello and Arthur Pecher

5.1 Introduction

In wave energy, perhaps more so than any other industry, the economics of product
development and product ownership are not separate from the product engineering
and design. This is the case because, despite high potential of untapped energy
resource and the constant attention of academic research and innovative companies
and inventors, as yet no one has verifiably achieved a minimum viable product in a
wave energy conversion system.

If a minimum viable product had been achieved by now then our task would be
simpler than it is. Evolutionary improvement due to incremental developments by
many individual subject experts would naturally follow any viable product.
Revolutionary leaps forward would also be easier to finance in the knowledge of an
already viable market. However, not for the want of trying, this is not currently the
status of wave energy research, and therefore a new more holistic approach is
needed.

Wave energy conversion systems are relatively complex systems and product
development is necessarily multidisciplinary. The evidence of wave energy
development experience so far is that excellence in each component discipline is a
necessary but not sufficient condition for development of a successful product. In
other words, it is possible that a programme that achieves excellence in each
individual discipline might still not achieve a viable product. A more holistic
approach that focuses on the big picture economics is needed.

R. Costello (✉)
Wave Venture Ltd., Unit 6, Penstraze Business Centre, Truro, TR4 8PN, UK
e-mail: ronan@wave-venture.com

A. Pecher
Department of Civil Engineering, Aalborg University,
Thomas Manns Vej 23, 9220 Aalborg Ø, Denmark
e-mail: apecher@gmail.com

© The Author(s) 2017
A. Pecher and J.P. Kofoed (eds.), *Handbook of Ocean Wave Energy*,
Ocean Engineering & Oceanography 7, DOI 10.1007/978-3-319-39889-1_5

The discipline of Systems Engineering provides a suitable framework for the holistic approach that might allow progress towards a viable wave energy conversion system. A definition of systems engineering is also an excellent introduction to the role of economic analysis in wave energy research and development:

> "the Systems Engineering process aims to assure the adequacy and completeness of the system for the customers' requirements while also balancing these objectives with available resources and the schedule of the system development programme."

Economic analysis is invoked twice in this definition, first in the customers' requirements which will logically include a requirement for a profitable electricity generation system, and second in the reference to available resources of the system development programme. Allocation of these scarce resources to alternative designs and alternative research programmes should be based on economic analysis.

This chapter introduces the methods of economic analysis that are relevant to wave energy in the hope that they will be applied by the technology development teams to optimise the next generation of wave energy converters and deliver a minimum viable product in a wave energy conversion system.

5.2 Power Is Vanity—Energy Is Sanity

The product of an electricity generating business is energy, electrical energy to be precise. The reason to risk stating the obvious is the need to emphasise that for an electricity generating business all other things besides electrical energy are not generally saleable products. In particular, power and energy, while obviously related, are not the same thing. **Energy** is the ability to do work and is measured in kilowatt-hour, (kWh) or megawatt-hour, (MWh).[1] **Power** is the instantaneous rate of transfer of energy and is measured in kilowatt (kW) or megawatt (MW) (see Footnote 1). The units that are sold are units of energy not power. The annual revenue of an electricity generation business is directly proportional to its annual energy production and strictly not directly related to its power capacity.

Annual Energy Production is simply the total energy produced over a one year period.

Annual Average Power is the average power over one year

$$Average\,Power\,[\text{MW}] = \frac{Energy\,\text{Production}\,[\text{MWh}]}{Time\,[\text{hours}]} \tag{5.1}$$

$$Annual\,Average\,Power\,[\text{MW}] = \frac{Annual\,Energy\,\text{Production}\,[\text{MWh}]}{24 \times 365\,[\text{h}]} \tag{5.2}$$

---

[1]The standard International System of Units (SI) units for power and energy are the Watt (W) and Joule (J). A Joule (J) is equivalent to a Watt × Second (Ws). More conventional units used in utility scale electricity are kW = 1000 W and kWh = 3 600 000 Ws.

Rated Power Capacity is the maximum power that can be generated over a sustained timeframe, say one or more hours, without damaging or overheating the equipment. Installed power capacity is, for most intents and purposes, the same as rated power capacity.

Capacity Factor of a generator is the ratio of its Average Power to its Rated Power Capacity

$$Capacity\ Factor = \frac{Annual\ Average\ Power}{Rated\ Power\ Capacity}$$
$$= \frac{Annual\ Energy\ Production}{24 \times 365 \times Rated\ Capacity} \tag{5.3}$$

An important input to the economic calculations in the following sections is the annual energy productivity. Understanding the relationship between the rated capacity and the annual energy yield is important. The relationship can be written using the capacity factor

$$Annual\ Energy\ Production = 24 \times 365 \times Capacity\ Factor$$
$$\times Rated\ Power\ Capacity \tag{5.4}$$

It should be obvious from the preceding equation that rated power capacity alone is insufficient information to estimate the energy productivity (or revenue) of an electricity generating business, capacity factor is also needed. Power capacity alone is the figure that is invariably publicised in media reports and in company publicity. However, a rated power capacity is meaningless unless it is accompanied by a capacity factor because both measures are needed to calculate annual energy productivity—"Power is Vanity—Energy is Sanity".

5.3 Economic Decision Making

This section will give a top down look at investment metrics without dwelling on the details of the inputs, later sections will discuss the wave energy specific details of the inputs (mainly costs, energy production and revenue) to these investment calculations. Discounted cash flow techniques are the state of the art in economic appraisal and analysis of investments. Several economic decision metrics use discounted cash flow including Net Present Value (NPV) and Levelised Cost of Energy (LCoE). NPV is the most universally applied measure of investability across all sectors of investment and LCoE is a widely used measure in electricity generation investment. These are discussed in the following sections along with a number of other relevant measures of investability.

Often companies or investors do not chose to invest on the basis of one criterion, but will evaluate the project using two or more criteria. Ranking of alternatives has to be based on a single metric, usually NPV, but additional metrics may be used as criteria for filtering projects that do not meet certain requirements. As a result it may be necessary to evaluate more than one of the measures presented in the following sections.

5.3.1 Cash Flow Terminology

Capital Expenditure (CapEx) is the total initial costs of setting up a project. In wave energy this includes; project planning and purchasing, transporting, installing and commissioning WEC's in a wave farm. Sometimes project planning and financing is called development expenditure and is separated from CapEx as DevEx but in the equations later in this chapter DevEx is treated as being included in CapEx.

Operational Expenditure (OpEx) is the ongoing annual cost of owning and operating a project, including all costs and payments except Taxes.

Decommissioning (Dec) is the costs of uninstalling and removing equipment after the useful life of the wave farm has been expended.

Revenue is the product of units delivered and sale price

$$Revenue = Annual\,Energy\,Production \times Power\,Purchase\,Price \qquad (5.5)$$

Operating Profit is the revenue less the OpEx

$$Operating\,Profit = Revenue - OpEx \qquad (5.6)$$

Tax on profits less allowable deductions is due to be paid to government. Tax is a cost and must be included in the cash flow analysis. Depreciation, or capital allowance, is usually an important allowable tax deduction, especially so in wave energy since the cost of equipment is so important. In some countries tax credits (production tax credits, installation tax credits) are an important strategic incentive mechanism.

$$Tax = Tax\,Rate \times (Revenue - OpEx - Tax\,Deductions) - Tax\,Credits \qquad (5.7)$$

Depreciation is not a cash flow but must be considered in detailed cash flow analysis because depreciation (or related concepts such as capital allowances) is usually an allowable tax deduction and so even though it is not a cash flow in itself depreciation can affect taxation which, unfortunately, is very much a real cash flow.

Cash Flow is the actual cash flow generated by the operations. (Some handbooks refer to this as Net Operating Profit Less Adjusted Taxes or NOPLAT)

$$Cash\,Flow = Revenue - OpEx - Tax \qquad (5.8)$$

Free Cash Flow (FCF) is the Cash Flow less the CapEx, it is a measure of the cash available in any time interval in the project lifetime.

$$FCF = Cash\,Flow - CapEx \tag{5.9}$$

Conventional Cash Flow is a common pattern of cash flows in a project. In a conventional cash flow the FCF will be strongly negative in the early years of a project due to the timing of CapEx, in later years as the project progresses the CapEx ends, the revenue is more significant and the FCF goes positive.

5.3.2 Time Value of Money (and Energy)

The expectation of earning interest on money deposited in the bank is commonplace. Another way of expressing this expectation is to say that the future value of the deposit will be greater than its present value. It is also normal to expect that this difference in value increases with the length of time that the investor has to wait for their returns. This relationship between future value (*FV*) and present value (*PV*) can be represented by the compound interest formula

$$FV = PV \times (1+i)^{n} \tag{5.10}$$

where i is the interest rate and n is the number of compounding periods, (the compounding period is often one year). Figure 5.1 shows the increasing path from present value to future value, if amount X is put on deposit its future value after twenty years is Y.

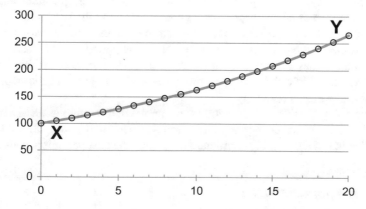

Fig. 5.1 Compound interest on deposit X at 5 % interest yields amount Y after 20 years. Or equivalently, if a future payment of Y is expected 20 years from now, it is equivalent to a payment of X now since X could be put on deposit now to get the same eventual payment

Since future value is greater than present value it follows that present value is less than future value, or in other words the present value of a future payment is less than the amount of the payment. So Fig. 5.1 can also represent that the future payment Y is equivalent to a payment X at the present time. The process of calculating the present value of future payments can be represented by the formula

$$PV = \frac{FV}{(1+d)^n} \tag{5.11}$$

where d is the discount rate. In the rest of the chapter we will use the notation $PV(X)$ to mean the present value of a future cash flow X.

$$PV(X) = \frac{X}{(1+d)^n} \tag{5.12}$$

The process of determining future value from the present value is called *compounding* and the opposite process of determining present value from the future value is called *discounting*.

In the formulas presented above the similarity between interest rates and discount rates is clear but in practice the terms are not interchangeable. As is common experience, interest rates generally apply to bank products such as savings, loans and mortgages. In most countries an official base rate of interest is set by a central bank. Discount rates on the other hand are used in assessing investments, especially investments in infrastructure projects. The central bank does not set a standard discount rate, each investor must choose an appropriate discount rate for each type of project. Discount rates commonly range from a similar level to interest rates up to significantly higher than interest rates.

Interest and discount rates are both intended to compensate an investor for the time waiting for the future payment and for the risk that the payment might not occur. In the case of a bank deposit or a government bond the risk of not receiving your money with interest is extremely low so the interest rate almost wholly represents compensation to the investor/depositor for the period of time that they must wait for their money to be repaid. In the case of future cash flows within a project the risk varies widely depending on the type of the project and the appropriate discount rate varies accordingly.

As implied by the title of this section the principles of discounting can be applied to productivity, in our case energy, as well as money. An implicit assumption underlying the application of discounting to productivity is that the cost per unit is constant.

5.3.3 Economic Metrics

Possible decision making metrics for use in wave energy projects are listed in Table 5.1. These are listed *approximately* in order of increasing sophistication. The first sub-group relate to energy generating projects are all measures of energy

Table 5.1 Selected economic metrics for wave energy technologies and wave energy projects

	Yield included	CapEx included	OpEx included	Revenue included	Time value of money	Relative/absolute	Defined for project Type	Notes
A. Productivity:								
Annual energy production (AEP)	Y	N	N	N	N	A	Energy	Limited applicability
Capture width	Y	N	N	N	N	A	Wave	Limited applicability
Capture width ratio	Y	S	N	N	N	R	Wave	Limited applicability
B. Cost of energy:								
AEP per unit displacement	Y	S	N	N	N	R	Wave	Limited applicability
AEP per unit surface AREA	Y	S	N	N	N	R	Wave	Limited applicability
AEP per unit CapEx	Y	S	N	N	N	R	Energy	Suitable for technology optimisation in early stage R&D
Levelised Cost of Energy (LCoE)	Y	Y	Y	S	Y	R	Energy	Recommended for choosing between energy projects and for technology optimisation in R&D. Best option if power purchase price is unknown
C. Investability:								
Payback period (PP)	Y	Y	Y	Y	N	R	All	Lower importance than NPV
Discounted payback period (DPP)	Y	Y	Y	Y	Y	R	All	Lower importance than NPV
Internal rate of return (IRR)	Y	Y	Y	Y	Y	A	All	Widely used and acceptable for all choices
Profitability index (PI)	Y	Y	Y	Y	Y	R	All	Recommended as complimentary to NPV
Net present value (NPV)	Y	Y	Y	Y	Y	A	All	Recommended for all choices

Approximately in order of increasing sophistication. *S* surrogate included in metric. *R* Relative measure. *A* Absolute measure productivity. In this first group "capture width" and "capture width ratio" are wave energy specific and have some additional limitations; they are usually calculated for a single device rather than a wave farm and are usually calculated for a single sea-state or regular wave rather than annual or multi-annual wave data. The difficulty with all measures in this first group is that they ignore both the cost and revenue components of a wave energy project and, for this reason, are not reliable decision metrics when taken alone. In some cases these metrics are intermediate results that are in any case needed to calculate the more advanced metrics and in other cases are trivial to calculate so should always be available to the analysis for comparison.

The second sub-group in Table 5.1 attempts to address this deficiency in the first group by including costs or surrogates for costs such as cubic displacement or surface area of the machinery. These surrogates are reasonable since the size of the equipment is an important driver of the capital cost of a wave farm, but metrics that use surrogates for actual costs are still not reliable decision metrics when used on their own. Cost of Energy and Levelised Cost of Energy (LCoE) include all cost data and are the most reliable metrics in this group. **Levelised Cost of Energy (LCoE)** is a cost of energy calculation that takes into account the time value of money. For energy generating projects LCoE is a valid decision making metric in its own right, and significant effort by the Carbon Trust, MARINET, NREL, IEA, and many others has been expended on defining procedures for calculating the LCoE for renewable energy projects. The third sub group in Table 5.1 are universal investment metrics that allow investment in wave energy to be compared to investment in any alternative project.

The remainder of this subsection presents a summary of concepts selected from Table 5.1.

AEP per unit CapEx, AEP per unit displacement and AEP per unit surface area

Annual Energy Production (AEP) per unit CapEx is a measure of economic performance that is limited principally by the fact that it neglects operating costs. AEP per unit displacement and AEP per unit surface area are similar measures that also neglect OpEx but, in addition, use displacement and surface area respectively as surrogates for CapEx. For some very large devices these surrogates may be well correlated with the device structural cost, and so are most relevant where the structural cost strongly outweighs the cost of other equipment such as PTO equipment. This argument is weakened, however, by the fact that the device structural cost sometimes makes up less (sometimes significantly less) than 50 % of the total CapEx and the CapEx due to balance of system might be much less well correlated with these surrogates than the structural cost. These metrics may be suitable for economic analysis very early in the R&D process, when insufficient information is available for more complete analysis, for example in choosing between design alternatives or concept alternatives. However, these are not sufficiently complete to be used for analysis to support project development decisions or device purchasing decisions.

Levelised Cost of Energy (LCoE) is a cost of energy calculation that takes into account the time value of money. For many analysts this is the most important measure of an energy investment. Many organisations have recommended specific methodologies and formulations for calculating the LCoE with various levels of sophistication that are appropriate for different applications. In general terms the LCoE is defined as

$$LCoE = \frac{Present\ Value\ of\ total\ costs\ over\ project\ lifetime}{Present\ Value\ of\ all\ energy\ over\ project\ lifetime} \qquad (5.13)$$

following this definition the equation for LCoE is

$$LCoE = \frac{\sum_{y=0}^{Y} PV\left(CapEx_y\right) + \sum_{y=0}^{Y} PV\left(OpEx_y\right) + \sum_{y=0}^{Y} PV\left(Dec_y\right)}{\sum_{y=0}^{Y} PV\left(AEP_y\right)} \qquad (5.14)$$

Equation (5.14) is similar to that given by the Carbon Trust's Marine Energy Challenge [1]. Renewable energy projects usually have conventional cash flow profiles, this means that the CapEx is always at the start and the revenue and OpEx are spread throughout the project. However in large wave farms it may not be possible to concentrate all the CapEx in a single year, or, for that matter, all the decommissioning in a single year either. Equation (5.14) is general in this regard; it does not make any assumptions about limiting any component of the cash flow to any particular time period. The equation for LCoE may be simplified if we give up some of this generality. If the costs of decommissioning are neglected and the CapEx is assumed to occur in the zero-th year then the equation becomes

$$LCoE = \frac{CapEx + \sum_{y=0}^{Y} PV\left(OpEx_y\right)}{\sum_{y=0}^{Y} PV\left(AEP_y\right)} \qquad (5.15)$$

A difficulty with LCoE (and all the previous metrics) is that it is only defined for energy projects, this is because LCoE uses annual energy productivity as a surrogate for revenue. LCoE actually ignores the market value of the energy product. Therefore it is only valid in comparisons between power generation options under comparable economic conditions and it should not be used to compare energy generating projects that would attract very different power purchase prices or tax rates, for example projects in different countries. A further limitation of LCoE is that it is not useful in comparing an investment in wave energy with any other investment opportunity outside the power generation sphere. In practice some investors may be specialised in energy, renewable energy or even in one particular type of renewable energy and are interested in choosing between power generating projects in a well understood market and regulatory regime, for these investors LCoE is a suitable choice of financial metric.

Case Study: SI-Ocean LCoE Methodology The Strategic Initiative for Ocean Energy (SI OCEAN) aims to provide a co-ordinated voice for the ocean energy industry in Europe and to deliver practical recommendations to remove barriers to market penetration. The following equation for LCoE is recommended in Ref. [37].

$$LCoE = \frac{SCI + SLD}{87.6 \times LF} \times \frac{d(1+d)^n}{(1+d)^n - 1} + \frac{OpEx}{87.6 \times LF}$$

where:

LCoE	Levelised cost of energy (€/MWh)
SCI	Specific Capital Investment (€/kW)
SLD	Specific Levelised decommissioning cost (€/kW)
	$\frac{SDC}{(1+d)^n}$
SDC	Specific decommissioning cost at end of lifetime (€/kW)
LF	Capacity Factor of wave farm [–]
d	Discount rate (%)
n	Operational life (years)
OpEx	Levelised O&M cost (€/kW/yr)

Source SI Ocean project, see Ref. [37].

Case Study: NREL onshore wind LCoE methodology The National Renewable Energy Lab in the US suggest calculating LCoE using a simplified formula designed to allow assessment of the true economic impacts of technical changes. The ICC, AOE and AEP input (defined below) characterise the technological performance (costs and output) the FCR input characterises the cost of financing.

$$LCoE = \frac{Present\ Value\ of\ total\ costs\ over\ project\ lifetime\ (\$)}{Present\ Value\ of\ all\ energy\ over\ project\ lifetime\ (MWh)}$$

$$LCoE = \frac{(FCR \times ICC) + AOE}{AEP_{net}}$$

where:

LCoE	Levelised cost of energy ($/MWh)
FCR	Fixed charge rate
	$\frac{d(1+d)^n}{(1+d)^n - 1} \times \frac{1 - (T \times PV_{dep})}{1 - T}$
ICC	Installed capital cost ($/kW)
AOE	Annual operating expenses ($/kW/yr)
	$LLC + O\&M(1 - T) + LRC$

d	Discount rate (%)
n	Operational life (years)
T	Effective tax rate (%)
PV_{dep}	Present value of depreciation (%)
CF_{net}	Net capacity factor (%)
LLC	Land lease cost ($/kW/yr)
$O\&M$	Levelised O&M cost ($/kW/yr)
LRC	Levelised replacement cost ($/kW/yr)
AEP_{net}	Annual Energy Production, net of losses and allowance for availability kWh/kW

Source NREL report, see Ref. [38].

Net Present Value (NPV) is the sum of the present values of the Free Cash Flow in all years of a project. NPV inherently accounts for the time value of money. The NPV tells us whether or not the present value of the operating profit is greater than the present value of the investment. The NPV is calculated from

$$NPV = \sum_{y=1}^{Y} \frac{FCF_y}{(1+d)^y} = \sum_{y=0}^{Y} PV(FCF_y) \qquad (5.16)$$

where d is the discount rate, FCF_y is the free cash flow in year y and Y is the lifetime of the project. The condition for investment is that the NPV should be strictly positive; projects with negative NPV are not investible while projects with positive NPV are investible. NPV is an absolute measure of performance, this means it gives the value of the investment rather than a ratio. See the Profitability Index later in this section for a relative measure that is complementary to NPV. The clarity around the decision making is one of the main advantages of NPV, however, it is partly illusory since the discount rate can be difficult to choose. See the section on weighted average cost of capital for methods used to set the discount rate. NPV is currently the most widely used and most reliable investment metric because choosing the projects with the highest NPV will maximise value which is, in principle, what best serves company shareholders.

Internal Rate of Return (IRR) is the discount rate that gives an NPV of exactly zero. It satisfies the following equation

$$0 = \sum_{y=1}^{Y} \frac{FCF_y}{(1+IRR)^y} \qquad (5.17)$$

The equation for IRR is implicit, it is most easily solved using a computer root finding algorithm, for example using the Newton-Raphson method. The equation for IRR is not guaranteed to have a single unique solution. In certain circumstances there may be no solution or there may be multiple solutions. Usually for projects with conventional cash flow there is either a single real solution or no solution. For a project with conventional cash flow no solution to the IRR equation may be interpreted as indicating an infeasible project. The uncertainty about the existence or uniqueness of the IRR makes it less suitable for use in automatic optimisation than LCoE or NPV.

5.3.4 Effect of Depreciation on Discounting

Depreciation is not a cash flow but must be considered in detailed discounted cash flow analysis because depreciation, or related capital allowances, is usually tax deductible. In practice it is advantageous to apply the highest rate of depreciation allowable under the applicable tax laws so that the benefits of the allowance are accumulated before they are eroded by inflation. For further information see Ref. [14].

5.3.5 Effect of Inflation on Discounting

The treatment of inflation can potentially make a difference to discounting calculations such as NPV, IRR, PI, DPP and LCoE. In a simplified assessment where tax allowances or even tax as a whole are neglected then inflation will make no difference but in a more detailed assessment care is required. A key concept related to inflation is constant and current euro (pound or dollar). Cash flow can be expressed in constant euro cash flow or current euro cash flow, $\overline{CF}_n$ and CF_n respectively. Current euro cash flow refers to the actual cash flow in year n, while the constant euro cash flow is the cash flow with the effects of inflation removed. The constant euro cash flow can be calculated from

$$\overline{CF}_n = \frac{CF_n}{(1+f)^n} \tag{5.18}$$

where f is the rate of inflation, assumed constant over n years.

When making a discounted cash flow assessment inflation can be included, current euro cash flow and nominal discount rate used, or inflation can be excluded, constant euro cash flow and real discount rate used. To calculate the real discount rate from the nominal and vice versa use

$$d_n = d_r + f + d_r f \tag{5.19}$$

when the terms d_r and f are small so that $d_r f \ll d_r$ and $d_r f \ll f$ then the equation may be approximated by

$$d_n \approx d_r + f \tag{5.20}$$

In the case without taxation NPV and other metrics are the same with and without inflation if the discount rate is adjusted to a real discount rate for the inflated cash flows. In the case with taxation the operating profit will be inflated along with all cash flows but the depreciation will not so the estimate of tax paid in current dollars will be higher when inflation is taken into account, accordingly the NPV will be lower when inflation is included. It is recommended to include inflation in assessments of well understood technologies for real projects and deployments. However a simplified approach is often justified at earlier stages. For example, in making a design choice in R&D between two alternatives it is unlikely that enough information will be available or well enough understood to allow the effect of inflation to have a reliable effect on the decision so the assessment should be simplified. For further information see Ref. [14].

5.3.6   Setting the Discount Rate

There are several methods for systematically choosing the discount rate. These include the weighted average cost of capital (WACC) and the risk adjusted discount rate (RADR). Companies may finance projects with a combination of equity, raised by selling shares to shareholders, and debt, borrowed from lenders. The Weighted Average Cost of Capital (WACC) also called the financial cost of capital is the weighted average of the cost of equity and the cost of debt. The equation for the WACC is

$$WACC = \left(\frac{E}{E+D}\right)i_e + \left(\frac{D}{E+D}\right)i_{dt} \tag{5.21}$$

where E is the equity amount, D is the debt amount (for a special purpose company with a single project $E+D$ is approximately equal to the total CapEx of the project), i_{dt} is the tax adjusted interest rate and i_e is the cost of equity. Equation (5.21) is sometimes modified for more than one type of equity each with a potentially different cost of equity. The tax adjusted interest rate is calculated from

$$i_{dt} = i(1-t) \tag{5.22}$$

where i is the interest rate and t is the tax rate. For an established company the cost of capital can be established by comparing historical returns to a market average using the Capital Asset Pricing Model (CAPM). Alternatively, the cost of equity may be calculated from a theory known as the dividend growth model

$$i_e = \frac{V_1}{P} + g \tag{5.23}$$

where V_1 is the expected dividend in the first year, P is the value of the company and g is the growth rate of the dividend.

Debt is generally cheaper than equity so a company will usually have a high debt to equity ratio, perhaps 4:1. However loan repayments are a fixed cost that makes a company vulnerable to interruptions in revenue so debt levels very much above this, called high leverage or gearing, may not be sound business practice. The WACC may be used directly in the discounted cash-flow calculations as the discount rate. Alternatively the Risk Adjusted Discount Rate (RADR) is

$$RADR = WACC + Project\ Risk\ Premium \qquad (5.24)$$

A survey of companies shows that most companies use the WACC as the discount rate and that most companies do not adjust the WACC for project risk i.e. the WACC is preferred over the RADR [14]. However, in a new industry such as wave energy, even though the WACC is likely to be higher than for other projects using more proven technology, a project risk premium is almost certainly appropriate.

The Carbon Trust's Marine Energy Challenge study [1] used discount rates in the range from 15 % for the first commercial wave energy devices to 8 % for wave energy when it is an established technology. The WaveNet European Commission Thematic Network [2] recommends a discount rate of 10 %, this is arrived at through use of the CAPM methodology. For comparison NREL recommendations for early (1995) onshore wind, in the absence of investment specific data, were rates of 3 % for government, 10 % for industry and 5 % for utilities [18].

As a closing point on selection of discount rate it is interesting to reflect on the implicit discount rate that individuals and households use when making non-business purchasing decisions. In general, consumers appear to apply much higher discount rates in their own lives than investors apply in infrastructure projects. Research by Hausman [8] and further research by Houston [9] found that households intuitively applied a discount rate of about 20 % when purchasing energy saving appliances. So it appears that private individuals can be more demanding investors than companies are.

5.3.7 Economic Decision Making—Which Metric to Use?

There are several types of decisions that should take economic assessments into consideration; these are not all in the deployment of large wave farms, some come much earlier in product development and R&D phase of a wave energy conversion technology. Selection of a metric to support decision making depends on the nature of the decision to be made and on the information available. The types of decisions that might be made with input from economic metrics include:

- Product development:

  - R&D management; allocate resources to competing sub-projects—which one will lead to a more competitive technology given available resources and timescale
  - Design decisions; choose between alternative design concepts- which one will lead to a more competitive technology with available resources and timescale

- Investment in WEC technology company:

  - Is the technology developed by the company competitive? The competitiveness of the technology is an important, if not the most important, component of the value of the company.

- Investment in wave-farm:

  - Is a particular wave-farm an attractive investment; on its own merits? Compared to other wave energy? Compared to other renewable energy? Compared to other electricity generation? Compared to any other investment?
  - Given a particular location is technology A or technology B more attractive?
  - Given a particular technology is location X or location Y more attractive? (Differences between location X and Y might not only be physical but may also be financial or political e.g. different energy prices, tax rates, insurance costs, permitting effort etc.)

Of key importance in determining which metric to use is the availability of the required input data. Critical, in this regard, is knowledge of the power purchase price. If the power purchase price is known then all of the metrics introduced in the previous sections are potentially available to the decision making process. If the power purchase price is not known then the revenue cannot be calculated and the LCoE is the most sophisticated economic metric that is available to decision makers.

In R&D management, especially in early stages of R&D, the decision is likely to be linked to a generic type of deployment location rather than a specific location with a known wave resource. It is also likely that the analysis should not be country specific but applicable to a wide range of markets. As a result the energy yield and the revenue are unknown or are subject to increased uncertainty and a cash flow based assessment is not appropriate, in this case a simplified LCoE assessment is recommended. It should be noted that in R&D and product development the immediate entrepreneurial goal might not be discovery of the technical configuration that will ultimately allow maximisation of NPV or IRR, it may instead be a minimum viable product.

In valuation of a company that produces wave energy conversion technology the competitiveness of the WEC technology is of critical importance. Similarly to the R&D decision making an assessment of a technology for company valuation should not be location specific or jurisdiction specific. It follows that LCoE is again an appropriate metric. Alternatively, the NPV could be calculated for a representative range of ocean locations and financial and regulatory environments.

In planning of large scale wave farm deployments the investment required is tens of millions of pounds or euro upwards. To attract this level of investment the project return must be attractive when compared to other investment opportunities that are available. If very large wave energy installations are to be privately financed then this will involve pension funds and other very large investment funds and these investors will compare wave energy to other investment opportunities outside the power generation sector. In this case NPV or IRR should be preferred over LCoE.

In principle the objective in investment decision making is maximisation of shareholder value. Crundwell (2008) notes that, maximising NPV maximises shareholder value while maximising PI maximises capital efficiency. If money and other resources were no object then it would be logical for all viable projects (NPV > 0. PI > 1) to proceed and in this case project assessment would be on project by project basis. However, in the real world resources and capital are constrained so making an investment decision is always done on the basis of ranking and choosing between alternatives. Even if only one project is proposed then in principle it should be compared to putting the investment amount on deposit.

In summary, LCoE is more likely to be independent of the financial/legal/taxation environment than NPV and conversely NPV is better able to reflect the effects of financial/legal/taxation issues than LCoE. If an assessment is technology focused then LCoE may be a better option than NPV. If the assessment is an investment focused on a specific deployment in a specific territory/location with known tariff/subsidy/tax/insurance conditions then NPV is a better choice than LCoE.

As final note, readers should be aware that while maximising NPV might maximise shareholder value, it is also true that NPV ignores external benefits (and potentially external costs) such as benefits of decarbonising electricity supply, reducing dependence on imported energy and other wider societal benefits such as providing employment [6]. An example of the need to take these wider benefits into account is the need for strategic government support for pioneering projects that allow projects with low NPV to proceed and facilitate learning that will drive costs down so that a new industry gains a foothold and projects with higher NPV and ever lower support requirements may follow.

5.3.8 Expert Oversight and Independent Review

The Electric Power Research Institute (EPRI) [4] correctly identify that it is possible to get almost any desired answer by making different assumptions. Similarly Stallard et al. [26] state that headline figures (e.g. €/kW or €/kWh) are useless unless the inputs and assumptions employed are clearly stated. It is therefore vital for WEC development companies to regularly receive an independent critique of their own projections of the cost of energy that their device might deliver. And for potential investors, customers and government sponsors to seek independent scrutiny of any estimates produced by a technology or project development company.

5.4   Economic Analysis in Technology R&D

Most energy utilisation is technologically intensive and all electrical energy generation and utilisation are technologically intensive. In the public consciousness energy and technology are often confused and the fact that energy and technology are not the same is often overlooked [22]. While energy is conserved, it is neither created nor destroyed. In contrast, the technology of energy conversion must be invented, researched, designed, manufactured, tested, refined etc. In other words research and development (R&D) is necessary. This section explores the importance of economic analysis in the innovation and R&D process.

Wave energy looks set to follow the industry structure of wind and solar PV energy, both have two intertwined businesses, one primarily concerned with energy and a second primarily concerned with technology.

- The *technology business* is concerned with the sale of energy conversion technology and the related activities of invention, research, development, design, demonstration and manufacture.
- The *energy business* is concerned with the sale of energy and the related activities of deploying, owning and operating the energy conversion technology and farm/facility.

(Each *business* is usually more than one company) Discussion of economics in renewable energy often focuses exclusively on the *energy business*, and economic analysis is usually focused on analysis to support project level go/no-go decisions. A tacit assumption underlying such discussion is that the energy conversion technology is available and mature. A second point that is ignored by focusing on the *energy business* is that R&D and other decision making within the *technology business* also needs to be supported by (very similar) economic analysis.

In wave energy some technologies have recently become available but are not yet mature. Wave energy economics must address the interlinked requirements of R&D in the technology business and project developments in the energy business. This link between the economics of these two businesses can be summarised as:

- Financing of R&D activities in the *technology business* relies on accurate and verifiable projections of attractive future project developments i.e. visibility of a future market for the technology.
- Project developments, and ultimate energy delivery, in the *energy business* rely on successful execution of product R&D and credible/verifiable analysis of technology performance.

5.5 Techno-Economic Assessment and Optimisation

The benefits of computer aided assessment of levelised cost of energy have long been recognised, for example the Carbon Trust and NREL both recommend Monte Carlo simulation as a tool for quantification of uncertainty in the LCoE. Farrell [7] and Dalton [13] separately demonstrate the use of Monte Carlo simulation in the economic assessment of wave energy projects.

Weber et al. [10] anticipates that techno-economic optimisation will form a crucial part of a successful performance before readiness product development. Effective implementations of integrated techno-economic optimisation have been demonstrated by [5, 16, 17] and this software is now becoming commercially available.

Figure 5.2 shows the structure of an integrated techno economic optimisation, courtesy of Wave Venture Ltd. The components of this particular integrated analysis are a physical model, an operational model, a cost model and a financial

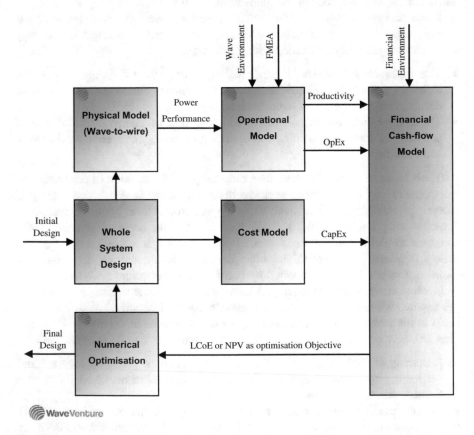

Fig. 5.2 Schematic of the information flow in an integrated techno-economic optimisation. A techno-economic assessment follows the same structure but without the numerical optimisation step which closes the loop. FMEA is Failure Modes Effects Analysis. Courtesy of Wave Venture Ltd

model. Part of the strength on the approach is its amenability to combine with numerical optimisation.

The physical model simulates the hydrodynamic interaction of the wave environment and the wave energy converter along with the performance of the devices power take off and power conversion chain, a so called wave-to-wire model. The input to the physical model is the system design and output is a characterisation of power performance and other engineering quantities of interest.

The operational model simulates the logistics of wave farm installation, operation and maintenance and ultimately decommissioning. The inputs of the operational model are the power characterisation calculated in the physical model, environmental data necessary to calculate weather windows and the system energy productivity and a characterisation of system reliability in the form of a failure mode effects analysis (FMEA). The outputs are estimates of the energy productivity and the operational expenditure. The advantages of this approach are that the availability and the operational expenditure are calculated by the simulation based on testable inputs instead of assuming an arbitrary percentage availability and an arbitrary operational cost based on experience in other sectors which might not relate to wave energy.

The cost model is formed from a suitable structure as introduced in the next section and is linked to the system design parameters so as quantities change the capital cost can be automatically updated.

The economic model generates a simulated discounted cash flow analysis which can be used to calculate any of the economic metrics introduced earlier in this chapter and potentially many more. A key advantage of the approach is that the economic value of a system design can be assessed without any third party data, especially third party performance data.

5.6 WEC Cost-of-Energy Estimation Based on Offshore Wind Energy Farm Experience

5.6.1 Introduction

Estimating the LCoE for a WEC array requires a lot of detailed information, which often can only be obtained after having completed several similar projects. However, valuable information on many of these specific topics can also be obtained by looking at the LCoE breakdown of offshore wind energy farms, which is now done in this section.

The structure of this LCoE calculations is following the document: "Value breakdown for the offshore wind sector" prepared by BVG Associates for the Renewables Advisory Board of the UK government [1]. The cost breakdown is done for a whole wind energy farm, not only for the wind energy technology itself. This document presents the relative cost of all main categories that are present in a 90 MW offshore wind farm in less than 20 m of water depth and a lifetime of 20 years, based on information provided by key industry players. These (sub-) categories and related cost, can be used to guide the cost of energy calculation for a similar WEC farm and help to estimate some of the sub-categories, which are too difficult/impossible to estimate with a reasonable accuracy for a WEC array at this point of time.

In addition, a presentation by Siemens Wind Power in late 2014 covering their actual LCoE and a related document by the International Renewable Energy Agency that covers the cost of renewable energy have been used [2, 14] to indicate reasonable level of costs of certain parameters.

5.6.2   Definition of the Categories

The definition of the (sub-) categories is taken directly from the value breakdown document (not everything has been reproduced here), and are thereby directly linked to an offshore wind energy project. The categories are as follow:

Development and consenting includes the multifaceted process of taking a wind farm from inception through to the point of financial close or commitment to build, depending on the contracting model, including Environmental Impact Assessment, planning, Front End Engineering Design studies and contract negotiation.

Turbine excluding tower includes supply of all components (including turbine transformers) upwards from (but excluding) the transition piece/foundation and in this case also excluding the tower structure. This includes delivery to a port (which may not be the port used for storage and pre-assembly of components before transfer to the wind farm site).

Balance of plant (BoP) includes detailed infrastructure design and supply of all parts of the wind farm except turbines, including tower, foundations, buildings, electrical systems between turbine and the onshore demarcation point between the farm and grid. Conventionally, the tower is seen as part of the scope of supply of the turbine. In this case, due to the synergies of manufacture of the tower and typical steel foundation, it has been incorporated here.

Installation and commissioning includes installation of turbines and balance of plant on site and commissioning of these to a fully operational state, up to point of issue of any take over certificate.

Operation and maintenance (O&M) starts from take-over, on completion of building and commissioning of all or part of a farm. It includes servicing of turbines and other parts including electrical grid connection. Whilst it does include insurance for the replacement of faulty/broken components or defective work it does not include coverage of this by warranties.

In addition, the following definitions are used:

Capital Expenditure (CapEx) includes all one-time expenditure associated with farm development, deployment and commissioning up to the point of issue of a takeover certificate.

Operating Expenditure (OpEx) includes all expenditure occurring from immediately after point of takeover, whether one-time or recurring, related to the wind farm, measured on an annual basis. Excluded are expenses inherent to the operation of the operators business but not directly related to the operation and management of the wind farm.

Grid connection includes the dedicated cables and other costs associated with connecting the farm to the National Grid, including any isolators and switchgear under the control of the onshore network operator.

Note that Project management, insurance and other costs relevant to many activities across the life of the farm have been included in these activities, rather than been separated out.

5.6.3 Wind Energy Project Case

5.6.3.1 Introduction

Some reference values need to be chosen, such as the kW price and the capacity factor of an offshore 3.6 MW wind turbine, as the relative cost of the sub-categories is given. The corresponding values depend on many factors (e.g. environmental resource and others), which is also reflected in the huge variation in their values that can be found in related literature. In order to give an example, an extract of the weighted average CapEx cost per kW provided by IRENA is given in Fig. 5.3.

By considering different sources in literature, under which a recent presentation by Siemens Wind Power [2], a CapEx price per MW of 4.5 m€ was chosen at a capacity factor of 30 %, a lifetime of 20 years and a discount rate of 10 %.

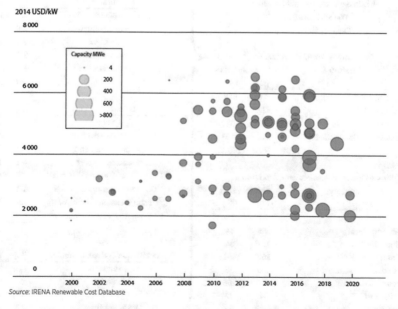

Fig. 5.3 Weighted average total investment for commissioned and proposed offshore wind energy projects 2000–2020, Courtesy of IRENA [14]

A ratio between CapEx and discounted OpEx of 73 % against 27 % is given by Siemens for a 1000 MW project of 6 MW turbines at 30 m of water depth. They also state a current LCoE of 0.145 € per kWh (in 2010) as the baseline for such kind of project [2]. This is quite high (much higher than is found in general literature), as it is for such a large farm with such large turbines, which should bring the cost down. Their additional cost, most likely arises from the additional water depth, which will be attempted to be taken into consideration as well.

5.6.3.2 Categories Cost Breakdown

Table 5.2 presents a typical cost breakdown of offshore wind turbines. The costs are divided over the main different cost categories, which was done following [1].

Table 5.2 Overview of the cost breakdown of a 3.6 MW offshore wind turbine [1]

Case specifications:

Turbine capacity [MW]	3.6			
Total installed capacity [MW]	90			
Amount of turbines	25.0			
CAPEX/MW [m€/MW]	**4.5**			
CAPEX/turbine [m€/turbine]	16			
yearly OPEX/MW [k€/MW/y]	**200**			
yearly OPEX/turbine [k€/turbine/y]	720			
=> discounted OPEX [m€/MW]	1.7			

lifetime of a windturbine:	20
Discount rate [%]:	10%
Capacity factor [%]	30%

CapEx breakdown/ Wind turbine

Category	Category cost [%]	[k€]
Development & Consent	4.0	648
Turbine - WEC	32.6	5281
Balance of plant	36.6	5929
Installation & Commisioning	25.7	4163
Total	98.9	16022

Annual OpEx cost breakdown / Wind turbine

Category	Category cost [%]	[k€]
Operation - remote	7.5	54
Operation - local	7.5	54
Maintenance - remote	11.6	84
Maintenance - local	26.6	192
Port activities - remote	8.0	58
Port activities - local	23	166
License fees	3.8	27
Other costs	12.0	86
Total	100.0	720

Development & Consent	[%]	[k€]
Environmental survey	0.3	49
Sea bed survey - Geophysical	0.1	16
Sea bed survey - Geotechnical	0.5	81
MetMast	0.3	49
Development services - engineering	0.9	146
Development services - other	1.9	308
Total	4.0	648

Turbine	[%]	[k€]
Blades	7	1134
Hub assembly	3.6	583
Gearbox	9.2	1490
Electrical system	8.2	1328
Other systems	4.6	745
Total	32.6	5281

Balance of Plant	[%]	[k€]
Tower	6	972
Foundations	16	2592
Cables - Inter array	1.4	227
Cables - Export	4.1	664
Offshore Substations - electrical	5	810
Offshore Substations - other	1.4	227
Onshore electrical - electrical	2	324
Onshore electrical - other	0.7	113
Total	36.6	5929

Installation and commisioning	[%]	[k€]
Foundations	7	1134
Cables	9	1458
Turbines	9	1458
Offshore substations	0.7	113
Total	25.7	4163

The general development, infrastructure and commissioning cost of a wind turbine in a project, thereby excluding the technology itself, is in the range of 7,2 million Euros, corresponding to about 45 % of the CapEx. This includes the development and consent, the installation and commissioning and a part of the balance of plant category (excluding the tower and foundations of the BoP).

5.6.3.3 Levelized Cost of Energy Estimation

The CapEx and OpEx were obtained following some assumptions on their cost and on the capacity factor, which were based on different sources of available literature. This cost breakdown was done for an offshore wind farm at 20 m of water depth. It seems to be more relevant to compare both case at 30 m of water depth, which correspond to the case of Siemens, which is stating a LCoE of 0.145 €/kWh. Therefore, an additional cost of 50 % was added to the tower and foundation (and its installation), to take this additional depth into account, which is (off course very simplistic but) considered to be conservative (Table 5.3).

Table 5.3 LCoE estimation for an offshore wind turbine

Cost		20 m water depth		30 m water depth	
			Ratio		Ratio
CapEx	(k€)	16022	72 %	16022	65 %
50 % extra cost on tower and foundation (part + installation)				2349	10 %
Discounted OpEx	(k€)	6130	28 %	6130	25 %
Total project cost	(k€)	22152		24501	
Revenue					
Power	(kW)	3600		3600	
Capacity factor	(%)	30 %		30 %	
Annual energy production	(MWh)	9467		9467	
Levelized cost of energy					
LCoE	(€/kWh)	0.117		0.129	

For the wind energy case at a water depth of 30 m a LCoE of 0.129 €/kWh is calculated. This seems to be reasonable when looking at most literature, however appears to be approx. 12 % lower then what Siemens estimates (LCoE of 0.145 €/kWh) for a much larger wind farm (1000 MW against 90 MW) and with larger turbines (6 MW against 3.6 MW). An additional factor of 50 % on the tower and foundation was maybe not sufficient, or it might maybe also affect other sub-categories which were not updated correspondingly (Fig. 5.4).

Fig. 5.4 Illustration of the
relative cost of the different
sub-categories of an offshore
wind turbine at 30 m of water
depth

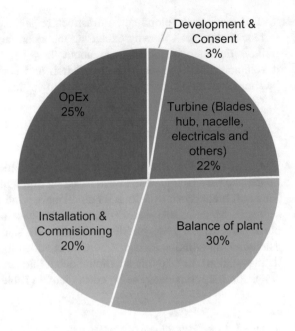

5.6.4 Wave Energy Case

5.6.4.1 Introduction

The same analysis can be done for a 90 MW WEC farm at 30 m water depth (or deeper). The same categories are maintained, with just some few adaptions into the sub-categories. The main adaptions, in order for it to fit the case of a floating WEC, are:

- The turbine category corresponds here to the WEC category. It aims at including the same scope, thereby excluding the mooring system.
- The mooring system is interpreted to correspond to the tower and foundation of the wind turbine and thereby put in the Balance of Plant category.

The resulting values relative to the WEC, have to come from a broad range of test and development efforts. The size and weight of the structure and sub components can be based on scaling, while the cost of materials and of components should be based on discussion with suppliers and quotations.

A cost breakdown of a 90 MW WEC array is made for two different sizes of WECs, 0.75 MW and 3.6 MW, based on the information of this offshore wind turbine case. The analysis aims to be generic and thereby no specific WEC technology is considered. The analysis assumes that the area required for a 90 MW array with types of WEC types is the same. General information regarding the cost of WECs can be found in e.g. [18].

5.6.4.2 Category: Development and Consent

The development and consent expenses of a wind energy farm are considered quite representative for a wave energy farm, as they both asses an offshore environment of somewhat the same specifications (water depth, project area, distance to shore and same objective to produce electricity). However, some of these expenses are to a certain extent dependent on the amount of WECs (how many detailed investigations need to be made e.g. on the soil) and on the technology (how detailed does some information need to be e.g. the seabed).

For a WEC array of 90 MW being deployed in about 30 m water depth, the overall survey costs are believed to be approximately the same, for large as well as for small WECs, as they will require approximately the same area, on the exception of the geotechnical sea bed survey. As this depends on the amount of systems to be installed (each one needs an analysis) and on the level of detail that is required (offshore wind requires much more detailed analysis). Therefore, the cost for small and large WECs have been reduced to 20 % relative to offshore wind value. All the other category costs have been divided amongst the amount of WECs that are required to make a 90 MW farm.

The development services cost are linked to the size of the project (same for all) and to some extend to the amount of systems to be installed (more cable routes, WEC positions and others to be analysed). Here, the same cost per large WEC as for wind turbines has been used, while for small WECs, the cost per WEC has been halved (Table 5.4).

Table 5.4 Overview of the development and consent costs (per unit) for a 90 MW farm of 25 × 3.6 MW WT and WECs and of 120 × 0.75 MW WECs

Category and sub-category	3.6 MW Wind (k€)	3.6 MW WEC (k€)	0.75 MW WEC (k€)
Survey			
Environmental survey	49	49	10.2
Sea bed survey-Geophysical	16	16	3.3
Sea bed survey-Geotechnical	81	20	20.0
MetMast	49	49	10.2
Development services			
Engineering	146	146	73.0
Other	308	308	154.0
Total cost	649	588	271

5.6.4.3 Category: Wave Energy Converter

This category corresponds to the main part of delivery by the wave energy developer, together with some few sub-categories in the Balance of Plant category, such as the mooring system. None of these values can thereby be taken from the offshore wind turbine case, as all of these are WEC technology dependant. The

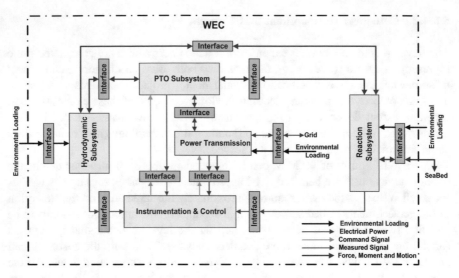

Fig. 5.5 Generic high level WEC design breakdown [9]

overall turbine category cost for a 3.6 MW offshore wind turbine (WT) has been estimated to be 5281 kEuro, corresponding to 33 % of the overall CapEx.

It is suggested to use the same sub-categories as proposed by DNV, which can be seen as a generic platform for the establishment of generic failure mode and effects analysis (FMEA) for WECs [8, 9] (Fig. 5.5).

All these categories contain different sub-categories dependant on the technology. A possible overview of this is given in Table 5.5.

Table 5.5 Overview of the possible cost breakdown for WECs

Category and sub-category	Weight or amount	Unit	Price/unit (k€)	Sub-category (k€)	Category cost (k€)
Reaction sub-system					
steel (painted, welded,…)	X	ton	X	X	
Concrete	X	ton	X	X	
Ballast	X	ton	X	X	
Others and extras				X	
				sub-total:	X
Hydrodynamic subsystem					
Steel	X	ton	X	X	
Concrete	X	ton	X	X	
Ballast	X	ton	X	X	
Others and extras				X	
				sub-total:	X

(continued)

Table 5.5 (continued)

Category and sub-category	Weight or amount	Unit	Price/unit (k€)	Sub-category (k€)	Category cost (k€)
PTO subsystem					
PTO unit	X	#	X	X	
Generator(s)	X	#	X	X	
Power electronics	X	#	X	X	
Others and extras				X	
				sub-total:	X
Instrumentation and control					
Cooling system	X	#	X	X	
Others (insulation, drain, wiring, ...)	X	#	X	X	
PLC-SCADA	X	#	X	X	
Instrumentation and communication	X	#	X	X	
Others and extras				X	
				sub-total:	X
R&D					X
Management					X
Contingency					X
Profit					X
				Total cost/WEC:	X

5.6.4.4   Category: Balance of Plant

The Balance of Plant (BoP) includes detailed infrastructure design and supply of all parts of the farm except for the WEC, including, foundations, buildings, electrical systems between WEC and the onshore demarcation point between the farm and grid.

Some of the costs here are very specific to the technology (mooring and foundations) and are thereby left blank for the WECs (noted with an "X"), while others can directly be taken over. The same cost for all the sub-categories has been maintained as for offshore wind (the total cost has been divided by the amount of WECs), except for the substation category, where the cost for small WECs is estimated to be a third than that of large WECs (although sharing the same platform, still requires more cable connections, routes and others) (Table 5.6).

Table 5.6 Overview of the balance of plants costs (per unit) for a 90 MW farm of 25 × 3.6 MW WT and WECs and of 120 × 0.75 MW WECs

Category and sub-category	3.6 MW wind (k€)	3.6 MW WEC (k€)	0.75 MW WEC (k€)
Tower/Mooring	972	X	X
Foundations	2592	X	X
Cables			
Inter array	227	227	47
Export	664	664	138
Offshore substation			
electrical	810	810	270
Other	227	227	47
Onshore electrical			
Electrical	324	324	68
Other	113	113	24
Total	5929	2365 + X	594 + X

5.6.4.5 Category: Installation and Commissioning

The cost of all the related sub-categories are case/technology dependant and can thereby not easily be derived from the wind energy project case. However, they can be used as inspirations and in the case where less work has to be performed at sea (more of the work can be done in the harbour); they can be assumed to be lower. Therefore, the overall cost for the installation of the cables and offshore substation is expected to be the same, while the work on the installation and commissioning of the foundations and WEC are expected to be significantly lower for WECs, thereby they have been reduced by 75 %.

However, you would expect that many of the costs would be roughly the same per all WECs, independently of the generator size, e.g. installation and commissioning of the WEC, electrical connections and installation of foundations etc. Thereby the cost per unit has been assumed to be 50 % lower for small WECs compared to large WECs, for all categories except for the substation (where the overall cost has been divided by the amount of WECs) (Table 5.7).

Table 5.7 Overview of the installation and commissioning costs (per unit) for a 90 MW farm of 25 × 3.6 MW WT and WECs and of 120 × 0.75 MW WECs

Category name	3.6 MW wind (k€)	3.6 MW WEC (k€)	0.75 MW WEC (k€)
Foundations	1134	284	142
WEC	1458	365	182
Cables	1458	1458	729
Offshore substation	113	113	24
Total cost	4163	2219	1077

5.6.4.6 Category: Operation and Maintenance (OpEx)

The operation and maintenance costs for large WECs are kept identical to the wind energy project, as there are various arguments that point in both directions. Some of the arguments in favour are that the WEC might be able to be decoupled and brought back to a harbour for maintenance, making large maintenance much easier. However, some parts of the device might be more difficult of access and there are more moving parts that are in contact with water (or being submerged). However, for WECs having most of their essential parts being submerged, the relative OpEx are expected to be much higher. For the WEC project, based on small WECs, it is expected that the relative OpEx will be significantly higher for several reasons, such as:

- The same effort (and thereby cost) is required to access or retrieve a large or a small WEC, this makes the relative cost higher for small WECs.
- The project made out of small WECs consists out of many more WECs (120 against 25). This means that in total many more sub-systems (each system requires a PTO, generator, mooring system, …) need to be serviced and maintained, which increases significantly the relative OpEx costs.

The OpEx cost for small WECs (with vital parts, such as PTO, not being submerged) is thereby assumed to be 50 % lower than that of large WECs, which is still assumed to be conservative (Table 5.8).

Table 5.8 Overview of the yearly operation and maintenance costs (per unit) for a 90 MW farm of 25 × 3.6 MW WT and WECs and of 120 × 0.75 MW WECs

Category and sub-category	3.6 MW Wind (k€)	3.6 MW WEC (k€)	0.75 MW WEC (k€)
Operation			
Remote	54	54	27
Local	54	54	27
Maintenance			
Remote	84	84	42
Local	192	192	96
Port activities			
Remote	58	58	29
Local	166	166	83
License fees	27	27	14
Other costs	86	86	43
Total annual cost	721	721	361

5.6.4.7 Overview and Levelized Cost of Energy Estimation

The mean annual energy production (MAEP), which is the multiplication of the capacity factor of the device times the installed capacity, is expected to be in the same range for a large WEC as for an offshore wind turbine. An average capacity factor (including the availability of the device) of 30 % has been assumed. This is expected to be significantly lower under certain circumstances, for small devices because their max-to-mean ratios of the absorbed power are much larger and their power smoothening capabilities are generally much lower. Their capacity factor (including availability) has thereby be assumed to be of 20 %, which is assumed to be reasonably conservative as a long-term projection.

In Table 5.9, an overview of the costs and energy production is given together with the LCoE. The total cost, corresponds to a "base" CapEx and discounted OpEx, while no specific cost for the WEC has been included (thereby marked by "X"). This base cost can also be set in terms of LCoE, and thereby represents the base electricity cost, not including the technology itself.

Table 5.9 Overview of the cost breakdown together with the base LCoE for a 90 MW farm of 25 × 3.6 MW WT and WECs and of 120 × 0.75 MW WECs

Costs	3.6 MW Wind (k€)	3.6 MW WEC (k€)	0.75 MW WEC (k€)
Development and consent	649	588	271
Turbine/WEC	5281	X	X
Balance of plant	5929	2365 + X	594 + X
Installation and commissioning	4163	2219	1077
Total CapEx	**16022**	**5822 + X**	**1941 + X**
Annual OpEx	721	721	361
Discounted OpEx (20 years)	**6138**	**6138**	**3069**
Total (CapEx & OpEx) costs	22160	11960 + X	5010 + X
Revenue			
Approx. capacity factor (%)	30	30	20
Mean annual energy production (MWh)	9467	9467	1315
Levelized cost of energy	(€/kWh)	(€/kWh)	(€/kWh)
Base LCoE (without OpEx)	0.085	**0.031 + X**	**0.074 + X**
Total LCoE	**0.117**	**0.063 + X**	**0.191 + X**

The total CapEx cost is composed of a "base" cost and a technology cost (marked by "X" for the WECs). This base cost is relatively independent of the technology, as it is mostly related to the project development, infrastructure and commissioning, while it is to some extent dependant on the generator size of the technology. The base cost is about 5.8 million Euros for a 3.6 MW WEC, while about a third of that for a 0.75 MW WEC. This corresponds to a base LCoE of 0.031 and 0.074 Euro/kWh for large and small WECs respectively. This means that the general development, infrastructure and commissioning costs weigh about 2.5 times higher on small than on large WECs.

When adding the OpEx cost to the base cost, the amount rises to 12 and 5 million Euros for the large and small WECs. This corresponds to a LCoE over the lifetime of the WEC, excluding the CapEx for the technology itself of 0.063 and 0.191 Euro/kWh for large and small WECs.

These results indicate clearly the economic advantage of large WECs. This is mainly because some of the costs are independent of the generator capacity of the WECs and that the capacity factor of a WEC usually increases with its physical size. It is thereby strongly beneficial to have a few large WECs instead of many small WECs in an array.

In order to be able to even further significantly reduce the base costs in the future, the scaling possibilities of a WEC technology are of very large importance.

5.6.5 Cost Reduction

Table 5.10 shows target costs for wave energy projects produced by Fitzgerald [33]. The table gives the OpEx in €m/MW/year and the CapEx in €m/MW that are necessary to give a 10 % IRR assuming a 160 €/MWh tariff. Different CapEx and OpEx values are given for a range of capacity factors (rows) and annual OpEx to CapEx ratios (columns).

Table 5.10 Affordable investment costs for generation projects [33]

Affordable investment costs for generation projects						
OpEx €m/MW/year			Annual OpEx as % of CapEx			
CapEx €m/MW			2 %	4 %	6 %	8 %
Capacity factor	20 %	OpEx	0.04	0.07	0.10	0.12
		CapEx	2.15	1.87	1.65	1.47
	25 %	OpEx	0.05	0.09	0.12	0.15
		CapEx	2.69	2.33	2.06	1.84
	30 %	OpEx	0.06	0.11	0.15	0.16
		CapEx	3.23	2.80	2.47	2.21
	35 %	OpEx	0.08	0.13	0.17	0.21
		CapEx	3.77	3.27	2.88	2.58
	40 %	OpEx	0.09	0.15	0.20	0.24
		CapEx	4.31	3.73	3.29	2.95
	45 %	OpEx	0.10	0.17	0.22	0.27
		CapEx	4.85	4.20	3.71	3.32

To yield a 10 % IRR for a 25 year Project life where a tariff or 160€/MWh is payable

The ratio of annual OpEx to CapEx increases column-wise from left to right and as a result the allowable CapEx to achieve the target IRR decreases from left to right. The capacity factor increases row-wise from top to bottom so that the allowable CapEx and OpEx to achieve the target IRR also increases from top to bottom. The annual energy yield and the project revenue can be expected to increase with the capacity factor.

In general costs may be expected to decrease as the number of units and the total installed capacity increases over time. This effect, known as the learning rate, was initially found to apply to aircraft and aerospace components and has since been confirmed to apply in many industries. Learning rates imply a pattern where each doubling of the capacity is accompanied by a consistent reduction in the unit price.

Figure 5.6, taken from the International Energy Agency report "Experience Curves for Energy Technology Policy" [34] shows the progress in cost of energy reductions as cumulative electricity production increased for a range of technologies. The percentages in braces in Fig. 5.6 are the "progress ratios", the ratio of price after to price before a doubling of capacity, e.g. wind power progress ratio is 82 %.

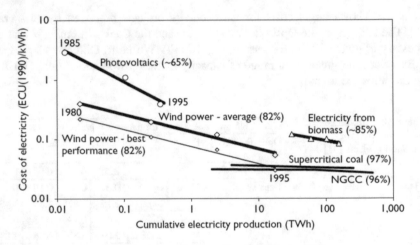

Fig. 5.6 Learning rates for different power generating technologies [34]

The learning curve theory does not propose any hypothesis for how the price reductions are actually achieved it treats the technology production system as a black box and only models an external view of the pattern of price over time. It is important to ask where price reductions might come from in wave energy.

Areas for further research in cost reduction in wave energy were investigated by the SI ocean project [39] and the recommendations include:

- Material optimisation
- Up-scaling of devices
- Batch and serial production
- Reduced levels of over engineering
- Improved moorings
- Improved foundations
- Cost effective anchors for all sea bed conditions

- Reduced cost of subsea electrical hubs/substations
- Optimisation of array electrical system and offshore grid
- Specialist installation vessels
- Improvements in weather forecasting
- Economic installation methods, e.g. fast deployment
- Improved ROV and autonomous vehicles

5.6.6 Revenue and Energy Yield

The final piece of the economics picture is annual revenue, which is directly proportional to annual energy yield. Other chapters in this book deal with wave resource characterisation and calculating and measuring the power absorption and power take off performance of WEDs in given wave conditions. This section will give only a brief discussion of the relation of these results to economic analysis.

Estimation of the annual energy yield must consider all of; wave resource, device power absorption performance, device power conversion/transmission efficiency and also availability. A key point in making an assessment of a wave energy project is that the energy sold to the electricity grid will be less than the energy absorbed by the wave energy device under continuous normal operation. The two principal reasons for this are firstly that there are losses involved in the power conversion and transmission steps that take power from the point of absorption to the point of metering and secondly that continuous operation of each device in the wave farm is unlikely. The implication of this is that a conservative assessment must allow for losses in the power take off and electrical power transmission and must also account for an availability that is less than 100 %.

Estimation of the annual revenue should consider annual energy production and the effective energy price including subsidisations and strategic supports, however the nature of subsidisation and strategic supports is varied and sometimes complex so the assessment may be as straightforward as calculating the product of annual productivity and effective price or it may be more involved. The next section will give a discussion of strategic support mechanisms.

5.7 Strategic Support Mechanisms

At any given time and place one form of electricity generation will provide cheaper electricity than all others. It stands to reason that all other forms of electricity generation are then more expensive or less attractive. If market forces alone decide investment in generation capacity then only power stations that use the most attractive technical solution will ever be built. Some form of market distortion or intervention is necessary to cause any technical solution other than the least expensive to be used. Reasons for making such an intervention include:

- Promotion of diversity of supply (and diversity related security of supply),
- Reduction of costs over a longer time horizon than considered by individual investors,
- Encouragement of (new) technologies with desirable characteristics
- Discouragement of (old) technologies with undesirable characteristics

Beyond energy related motivations policy related motivations[2] may include

- Protection of an established or domestic industry against encroachment of new or foreign industries
- Promotion or creation of a new or domestic industry in preference to older or foreign industries

Interventions are often targeted at influencing the decision to use a particular, already mature, technology, i.e. choice of technology at the pre-construction project planning stage, other interventions are targeted at increasing R&D investment in new technologies and a minority are targeted at influencing operational decisions e.g.: fuel-mix in co-firing or CHP operations management (see CHPQA). Interventions can take the form of regulations that discourage or effectively block a particular technology but more often interventions are structured as strategic support mechanisms that encourage a particular technology or behaviour.

Strategic support mechanisms maybe categorised as one of either **Market Pull** or **Technology Push**. Market pull is usually related to production incentives while technology push is related to either installation incentives or to research and development funding. Market pull type support mechanisms are effective in encouraging technology that is either already mature or can be made sufficiently mature with private investment, it is intended to heighten the price signal that activates private investment. Technology push, on the other hand, is effective in encouraging research in technologies that are not yet sufficiently close to commercialisation to benefit from price signals or market pull type supports. Technology push can also activate private investment through matched funding requirements.

Strategic supports whether market pull or technology push may take the form of

- Direct payments e.g.: feed in tariff, research grants, government contracts.
- Tax credits e.g.: production tax credit, installation tax credit, accelerated depreciation, R&D tax credit.

---

[2]In addition, but arguably less relevant to this discussion, a policy related motivation for intervention that is prevalent in some developing countries is reduction of consumer energy bills for welfare (or electoral) purposes. Such policies are a distortion that blocks price signals and discourages energy efficiency [35].

- In-kind or preferential provision of goods or infrastructure e.g.: access to materials, technology, natural resources, sea-bed lease, port construction, road construction.
- In-kind or preferential provision of services, especially services that transfer risk from investors to government e.g.: Loan guarantees, construction cost guarantees, demand guarantee, price regulation, market access regulations, favourable licensing and permitting.

Some governments have a philosophical objection to distorting free markets, but experience from studies of the nuclear industry and to a lesser extent the petrochemical industry has shown that this objection only leads to subsidies becoming hidden and more subtle. A consequence of the hidden nature of such subsidies is that they are sometimes so inscrutable that they can be denied. Proponents the nuclear energy often claim that nuclear energy receives no subsidisation when in fact it benefits massively from favourable long term power purchase agreements and from large scale transfers of risk and liability from the operators to the state [36].

Both market pull and technology push type strategic incentives are now needed to attract sufficient private finance to wave energy development. There are three key challenges that must be overcome by the wave energy industry and strategic incentives can play a role in addressing each, these key challenges are:

- Identify and develop those WEC concepts that are capable of reaching TRL9 i.e. have sufficiently high TPL and sufficiently lean/affordable development trajectory.
- Facilitate finance of the latter phases of development, demonstration and risk reduction (from TRL 6 to TRL8/9) where product development becomes too expensive for the SME's that typically initiate new and innovative technologies.
- Facilitate insurance against warranty claims after the start of volume sales.

Technology push type incentives and application of advanced R&D management techniques such as the Weber matrix as introduced in Chap. 4 will assist with the first of these challenges. A combination of market pull type incentives such as long term price supports, capital grants and crucially risk sharing such as loan guarantees and insurance initiatives such as government underwriting of project risk will assist with the second and third challenges.

References

1. BVG Associates for the Renewables Advisory Board.: Value breakdown for the offshore wind sector (2010)
2. Christensen, B.: Den nødvendige indsats i offshore vindindustrien—Siemens Wind Power. aarsmoedet 2014 i offshoreenergy.dk (2014)

3. Costello, R., Teillant, B., Weber, J., Ringwood, J.V.: Techno-economic optimisation for wave energy converters. In: International Conference on Ocean Energy, Dublin (2012)
4. Crundwell, F.K.: Finance for Engineers, Evaluation and Funding of Capital Projects. Springer, London (2008)
5. Farrell, Niall, O'Donoghue, Cathal, Morrissey, Karyn: Quantifying the uncertainty of wave energy conversion device cost for policy appraisal: An Irish case study. Energy Policy **78**, 62–77 (2015)
6. Future Marine Energy.: Results of the marine energy challenge: cost competitiveness and growth of wave and tidal stream energy. J Callaghan, R Boud—Carbon Trust (2006)
7. Fitzgerald, J.: Financing Ocean Energy Technology Development. ESBI 2012. http://www.ucc.ie/en/media/research/hmrc/forums/thirdforumoneconomicsofmarinerenewableenergy2012/JohnFitzgeraldOceanEnergyFinance-ESBPresentation.pdf
8. Hamedni, B., Ferreira, C.B.: Generic WEC risk ranking and failure mode analysis generic WEC risk ranking and failure mode analysis (2014)
9. Hamedni, B., Ferreira, C.B., Cocho, M.: Generic WEC system breakdown (2014)
10. Hardisty, P.E.: Environmental and Economic Sustainability, Environmental and Ecological Risk Assessment. CRC Press (2010)
11. Hausman, J.: Individual discount rates and the purchase and utilization of energy-using durables. Bell J. Econ. (1979)
12. Houston, D.A.: Implicit discount rates and the purchase of untried, energy-saving durable goods. J. Consumer Res. **10**. (1983)
13. International Energy Agency.: Experience Curves for Energy Technology Policy (2000)
14. International Renewable Energy Agency (IRENA).: Renewable power generation costs in 2014 (2015)
15. Koplow, D.: Nuclear Power: Still Not Viable without Subsidies. Union of Concerned Scientists (2011)
16. Kunstler, J.: The Long Emergency: Surviving the End of Oil, Climate Change, and Other Converging Catastrophes of the Twenty-First Century. ISBN-155584670X, 9781555846701. Grove/Atlantic, Inc. (2007)
17. Madlener, R., Ortlieb, C.: An Investigation of the Economic Viability of Wave Energy Technology: The Case of the Ocean Harvester," FCN Working Paper No. 9/2012, Institute for Future Energy Consumer Needs and Behavior, RWTH Aachen University (2012)
18. Ocean Energy Systems (OES).: International levelised cost of energy for ocean energy technologies (2015)
19. Ocean Energy: Cost of Energy and Cost Reduction Opportunities.: SI-Ocean Project. May 2013. http://si-ocean.eu/en/upload/docs/WP3/CoE%20report%203_2%20final.pdf
20. Padeletti, D., Costello, R., Ringwood, J.V.. A multi-body modelling approach for wave energy converters employing nonlinear joint representation. In: Proceedings of the OMAE, San Francisco (2014)
21. Previsic, M., Siddiqui, O., Bedard, R.: EPRI Global E2I Guideline. Economic Assessment Methodology for Offshore Wave Power Plants. E2I EPRI WP-US—002 Rev 4 (2004)
22. Short, W., Packey, D.J., Holt, T.: A manual for the economic evaluation of energy efficiency and renewable energy technologies. NREL/TP-462-5173 (1995)
23. Stallard, T., et al.: Economic assessment of marine energy schemes. In: Proceedings of the 8th European Wave and Tidal Energy (EWTEC) Conference, Uppsala, Sweden (2009)
24. Tegen, S., Hand, M., Maples, B., Lantz, E., Schwabe, P., Smith, A.: 2010 Cost of Wind Energy Review. NREL/TP-5000-52920 (2012)

25. Teillant, B., Costello, R., Weber, J., Ringwood, J.V.: Productivity and economic assessment of wave energy projects through operational simulations. Renew. Energy **48**, 220–230 (2012)
26. VV.AA.: WaveNet: Full report (2003). www.wave-energy.net/Library/WaveNetFullReport (11.1).pdf
27. Weber, J, Costello, R., Ringwood, J.V.: WEC technology performance levels (TPLs)—metric for successful development of economic WEC technology. In: Proceedings of the European Wave and Tidal Energy Conference (EWTEC), Aalborg (2013)

Chapter 6
Hydrodynamics of WECs

Jørgen Hals Todalshaug

6.1 Introduction

In this chapter we look at the fundamental principles of wave absorption, and of forces on floating bodies. The goal is to build an understanding of the main physical effects involved when trying to extract power from ocean waves.

6.1.1 Wave Energy Absorption is Wave Interference

Imagine a wave travelling in an open ocean area. It carries a certain amount of power. Then put a wave energy absorber in that area under influence of the same incident wave. If there is less energy travelling the ocean after you put the wave energy device there, it means that the device has absorbed energy!

Absorption of energy from gravity waves on the ocean follows the same basic principles as absorption of other types of waves such as electromagnetic waves (for instance radio and telecommunication) and sound. Wave energy absorption should primarily be understood as wave interference: In order to absorb a wave, the wave energy device must generate a "counter-wave" to interfere with the incident wave. If the interference is *destructive* (which in this context is positive!), such that the wave in the ocean is reduced, wave energy is absorbed by the device. This fundamental relation may be formulated in the following way: "A good wave absorber is a good wave maker" [1].

Figure 6.1 illustrates how the wave reflected at a vertical wall (left) may be cancelled by proper wave generation (right). In order to obtain cancellation by destructive

J.H. Todalshaug (✉)
Department of Marine Technology, Norwegian University of Science
and Technology (NTNU), Trondheim, Norway
e-mail: jorgen.hals@ntnu.no

J.H. Todalshaug
CorPower Ocean AB, Stockholm, Sweden

A. Pecher and J.P. Kofoed (eds.), *Handbook of Ocean Wave Energy*,
Ocean Engineering & Oceanography 7, DOI 10.1007/978-3-319-39889-1_6

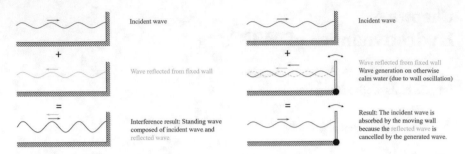

Fig. 6.1 Wave energy absorption as wave interference (Reproduced from lecture notes by Johannes Falnes and Jørgen Hals Todalshaug)

interference between the reflected and the generated wave, it is important that both the phase (timing) and the amplitude (strength) of the generated wave are chosen properly. This is crucial for any wave energy device, but the phase and amplitude may not always be easily controllable. See Sect. 6.1.8 for further treatment of this issue.

6.1.2 Hydrostatics: Buoyancy and Stability

Rigid-body motions are usually decomposed in six modes as illustrated in Fig. 6.2.

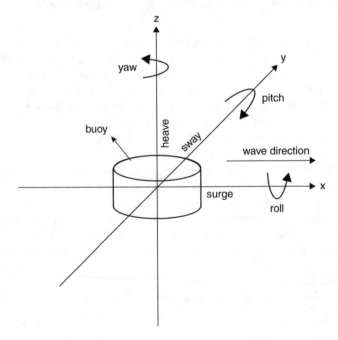

Fig. 6.2 The motion of a floating body is decomposed in translation: surge (1), sway (2) and heave (3); and rotation: roll (4), pitch (5) and yaw (6). The numbers in parenthesis are often used as index for each of these modes

In order to float steadily, a floating body must have large enough volume compared to the mass, and be hydrostatically stable for rotations in roll and pitch. It is hydrostatically stable if the sum of the gravity force and the buoyancy force gives a positive *righting moment*; a net moment working to bring the body back to hydrostatic equilibrium in case of disturbances, see Fig. 6.3. The stability may be characterised by a GZ curve, which gives the moment arm as function of the tilt angle α as shown in Fig. 6.4.

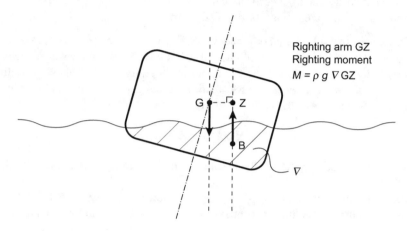

Fig. 6.3 Righting force $\rho g \nabla$

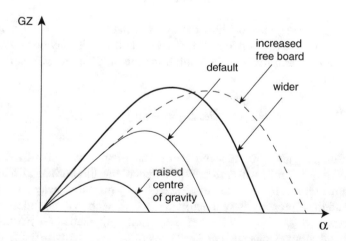

Fig. 6.4 GZ curve and how it is influenced by changes in geometry. Inspired by [2]

The buoyancy force on a partly or fully submerged body is found as

$$F_b = \rho g \nabla$$

the product of water density ρ, acceleration of gravity g and submerged volume ∇. The centre of buoyancy is found at the centroid of the submerged volume.

For small tilt angles, the righting moment depends on two factors:

- The vertical distance between the centre of mass and centre of buoyancy
- The water-plane area (both its size and distribution)

This means that a floating body may be made more stable by either lowering the centre of mass, raising the centre of buoyancy or by increasing the water-plane area. This will increase the slope of the GZ curve at tilt angle $\alpha = 0$. This slope defines the *hydrostatic stiffness coefficient* for rotation in roll or pitch, which is the increase in righting moment per change in tilt angle:

$$S_{4,5} = \rho g \nabla d GZ_{4,5}/d\alpha$$

If the body is axi-symmetric, the GZ curves for roll (mode 4) and pitch (mode 5) are equal.

The restoring force $F_3 = S_3 \eta_3$ for vertical motion η_3 may be defined by the hydrostatic stiffness coefficient in heave. It only depends on the water plane area A_{wp}:

$$S_3 = \rho g A_{wp}$$

Depending on the shape and how the axis of rotation is chosen, an excursion in heave might induce a rotation in roll or pitch (or both). This may be described by a stiffness coupling term that tells how large the pitching torque will be for a given heave excursion:

$$\tau_5 = S_{53}\eta_3 \text{ with } S_{53} = -\rho g \iint_{A_{wp}} x \cdot dS,$$

where the integral gives the first moment of area about the axis of rotation, and correspondingly for roll, with the only difference that the horizontal off-axis distance x (surge direction) needs to be replaced by $-y$ (negative sway direction). The rotational stiffness property is symmetric, $S_{53} = S_{35}$, such that the following is also then true: $\tau_3 = S_{35}\eta_5 = S_{53}\eta_5$. Further details on hydrostatics for floating bodies may be found in textbooks such as [3, 4].

6.1.3 Hydrodynamic Forces and Body Motions

Hydrodynamics is the theory about forces on and motion of fixed and floating bodies in moving fluids.

A wave energy converter typically experiences the following external loads:

- Gravity
- Buoyancy
- Excitation from incident and diffracted waves
- Wave radiation (forces due to generated waves)
- Machinery forces (PTO force incl. friction)
- Drag: Form drag and skin friction
- Wave drift forces
- Current forces
- Mooring forces

Its inertia governs how the floating body responds to these loads.

The effects of gravity and buoyancy are treated in the above section about hydrostatics.

In principle, the most important forces from the fluid are pressure forces arising due to incident waves and body motions. It is standard practice to divide these pressure forces in *excitation* and *radiation* forces based on a linearised (and thus simplified) description of the problem. Drift forces and form drag must then often be included as corrections in order to yield a sufficiently accurate description of the system behaviour. Skin friction may usually be neglected in wave energy problems.

Mathematically, the motion induced by the combination of these forces may be described by the following equation of motion,

$$(m + m_r(\omega))\ddot{\eta}_i + R_r(\omega)\dot{\eta}_i + S\eta_i = F_e + F_{PTO}$$

For simplicity, we have here assumed a regular wave input of angular frequency ω, motion in only one mode i, and disregarded drag, wave drift, current and moorings. The symbols in the equation refer to:

- η_i—position in mode i, with time derivatives $\dot{\eta}_i$ (velocity) and $\ddot{\eta}_i$ (acceleration)
- m—body mass
- $m_r = m_{r,ii}$—added mass
- $R_r = R_{r,ii}$—radiation damping (due to generated waves)
- $S = S_{ii}$—hydrostatic stiffness; the combined effect of gravity and buoyancy
- $F_e = F_{e,i}$—excitation force
- F_{PTO}—powertake-off (PTO) force, including losses

The equation above is that of a damped harmonic oscillator, such as a mass-spring-dashpot system, forced by an applied load. In our case the applied load is the combination of wave excitation and machinery load. In the following the different forces are explained more in detail without going into mathematical descriptions. Extensive treatment of hydrodynamic theory may be found in textbooks such as [3, 5, 6].

6.1.3.1 Excitation and Radiation Forces—Added Mass and Radiation Resistance

Keeping drag forces apart, the excitation forces are those felt by the body when kept fixed in incoming waves, whereas radiation forces are those felt by the body when moved in otherwise calm water.

The excitation force is found from the hydrodynamic pressure in the sum of incident and scattered waves. If the body is small compared to the wave length, scattering may sometimes be neglected, such that a rough approximation of the excitation force may be found considering only the pressure in the undisturbed incident wave. An improved approximation may be found by use of the so-called *small-body approximation,* or other, that includes a simplified representation of the force produced by the wave scattering.

Forces arising due to body motions are usually referred to as radiation forces. It is common to divide these forces in added-mass forces, proportional to body acceleration, and wave damping forces, proportional to body velocity. The wave damping may also be referred to as radiation resistance.

Physically, the added mass force may be pictured as an inertia force relating to the mass of water entrained with the body motion. It is important to realise that we do not speak of a fixed amount of water—In principle, all of the water is influenced by the motion of a floating body. The added mass coefficient is rather an equivalent quantity telling how large the fluid inertia force becomes when the body is accelerated. When averaged over time, there is no net power flow between the body and the fluid due to the added mass force.

The radiation resistance (or wave damping) force, on the other hand, is closely linked to the average power exchanged between the sea and the body. This force arises due to outgoing waves generated when the body moves. The radiation resistance coefficient tells how large the waves will be. As these are the waves that interfere with the incoming waves, the radiation resistance also indirectly tells how much power we can extract from the incoming waves. As you may understand, this makes the radiation resistance a very important parameter for wave energy extraction.

Unfortunately, in hydrodynamics, both the added mass coefficient and the radiation resistance coefficient depend on the frequency of oscillation, as indicated by the frequency argument in the equation of motion given above. This makes both modelling and optimisation of converters more challenging than it would otherwise be. The optimal device parameters and machinery settings depend on the wave frequency, which may be constantly changing!

There is a relation between the radiation damping and excitation forces, as both are measures of how strongly linked the body is to the wave field at sea: A body that is able to radiate waves in one direction when moved will experience excitation when acted upon by incident waves coming *from* that direction.

6.1.3.2 Machinery Forces

The machinery (PTO) forces are what separate wave energy converting systems from conventional marine structures. Hydrodynamically it does not matter how the machinery forces are produced. The force must be applied between the wave-absorbing body and a body fixed to the shore or to the sea bottom, or alternatively it may be applied between two floating bodies.

A very common assumption is that the machinery behaves like a linear damper where the force is proportional to velocity. This makes mathematical modelling easy, but there is, however, no general advantage in doing so in practice. What matters is how much and when in the oscillation cycle energy is extracted from the mechanical system by the machinery. See Sect. 6.1.8 for further details.

6.1.3.3 Drag Forces

Drag forces on a floating body mainly stem from vortex shedding when water flows past the body surface, or past mooring lines or other submerged parts of the system. As such, the drag force originates from a loss of kinetic energy. In general, the form drag forces increase quadratically with the relative flow velocity between body and water. If a wave energy converter is made so as to avoid drag losses when operating in normal-sized waves, as suggested in the section on design below, drag forces may still become important in high sea states and storm conditions due to this quadratic relationship.

The scaling of drag forces between model scale and prototype scale is not straight-forward. The flow regime depends on the scale parameter, such that geometric scaling of drag forces may not be applied directly when translating between small-scale experiments and full-scale testing. If, however, it can be established that the drag forces are due to vortex shedding around corners and they are also of secondary importance relative to other loads, geometric scaling may be expected to be a good approximation for the overall system [7].

6.1.3.4 Wave Drift, Current and Mooring Forces

Finally, we have forces that usually give the effect of slow low-frequent excitation and response, namely wave drift forces and mooring forces.

Wave drift forces are due to non-symmetric wave loading on bodies, to inter-action between waves of different wave period and to interaction between the wave oscillation and oscillation of the body itself. These effects give a constant or low-frequent excitation of the system. This excitation may become important if the waves are large, or if its period of oscillation coincides with resonance frequencies introduced by the mooring system.

In some locations, tidal and ocean currents give significant forces on the wave energy converters.

Slack-line mooring systems are usually designed to provide a positioning force to counteract the horizontal wave drift and current forces, whereas taut-line systems may additionally counteract wave-frequent excitation in one or more modes. Unless the mooring lines are used as force reference for the energy conversion, the mooring system is usually designed to give as little influence on the wave absorption process as possible. Low-frequent resonance in the mooring system may be detrimental for the lines in storm conditions or even in normal operating conditions if not properly designed. (See Chap. 7 on mooring)

6.1.4 Resonance

Resonance occurs when a system is forced with an oscillation period close or equal to the system's own natural period of oscillation. Such a resonance period exists if the system has both stiffness and inertia. Thus, for freely floating bodies we have resonance periods for heave, roll and pitch modes. With mooring lines connected we may in addition have stiffness in surge, sway and yaw, giving rise to resonance also in these modes of motion.

The analysis of harmonic oscillators such as described by the equation of motion above tells us that the system will resonate at the period of oscillation where the reactance of the system is zero, $\omega_0(m+m_r) - \frac{S}{\omega_0} = 0$. This happens when the potential energy storage and the kinetic energy storage of the system are of equal size [5]. Solving this equation gives a resonance period of

$$T_0 = 2\pi/\omega_0 = 2\pi\sqrt{(m+m_r)/S}$$

The relative bandwidth of the system is given by

$$\frac{\Delta\omega_{res}}{\omega_0} = \frac{R_r}{\sqrt{S(m+m_r)}}$$

For modes of motion with no or low stiffness ($S \to 0$) the relative bandwidth automatically becomes very large. The relative bandwidth is a measure of how strongly the system responds to inputs of frequencies other than the resonance period. Because ocean waves come with varying wave frequency, this is an important property for a wave energy converter. We will return to this subject later.

A useful approximation for the heave resonance period of a freely floating body may be derived if we assume that the cross-section of the buoy is fairly round and relatively constant with depth, such that the heave added mass m_r may be estimated from the width d of the body:

$$T_{0,3} \approx 2\pi\sqrt{(l+d/3)/g}$$

Here l is the draft of the freely floating body.

When a system is resonating with the incident waves, it means that motions tend to be amplified, resulting in large accelerations and forces. For this reason resonance is usually avoided in conventional naval architecture. A wave energy converter, on the other hand, may have to operate at or close to resonance in order to obtain a sufficient power conversion, which is dictated by the phase alignment between excitation force and body velocity. At resonance these are aligned. See also Sect. 6.1.8.

6.1.5   Oscillating Water Columns—Comments on Resonance Properties and Modelling

A simplified model to understand the dynamics of the oscillation water column (OWC) would be to think of it as an internal oscillating body of mass equal to the mass of the water in the column. In analogy with the expression for heave buoys, the resonance period may then be estimated roughly as

$$T_{0,owc} \approx 2\pi\sqrt{l_c/g}$$

assuming that the cross-sectional area of the column is fairly constant, and where l_c is measured along the centerline of the column from the inlet to the internal free surface as illustrated in Fig. 6.5. If the column is inclined, as in Fig. 6.6, the denominator must be replaced by $g\cos(\theta)$, where θ is the slope angle relative to the vertical. The absorption bandwidth of water columns may be increased by making a harbour-like construction at the inlet, with side walls reaching out towards the sea [8].

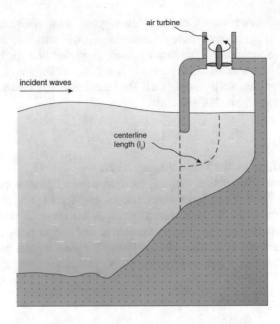

Fig. 6.5 Oscillating water column where the length l_c is measured along the centreline

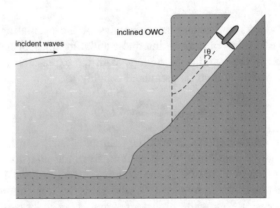

Fig. 6.6 Oscillating water column inclined at an angle θ

For mathematical modelling of OWCs, a simplified model could be established by applying the same rigid-body representation as used for floating bodies, where the free surface of the water column is thought of as a rigid lid [9]. A more accurate representation could be made by taking the free surface and chamber pressure into account [5, 10].

For oscillating water columns (OWCs), the chamber gas volume gives a compressibility effect that may be important. In model tests of OWC devices, it is crucial to remember that this compressibility does not scale geometrically: In model scale the chamber must be represented by an increased volume in order to give a representative stiffness force. The scaling factor for the chamber volume should be taken as N_L^2, where N_L is the length scaling factor. Thus for a model at 1:10 scale the chamber volume should be made equal to $1/100$ of the full-scale volume, rather than $1/1000$. Although scaled, the compressibility effect of the model will only be correct for small column excursions.

6.1.6 Hydrodynamic Design of a Wave Energy Converter

From the above discussion on waves, wave excitation and radiation we find that, in order to be suitable for absorbing waves, a body must have a shape, size, placement and motion that gives considerable outgoing waves when moved. In the following we will discuss general guidelines for wave energy converter design.

6.1.6.1 Size and Shape

The first rule of thumb for wave-absorbing bodies and water columns is that corners should be rounded. Sharp edges will induce drag and viscous losses that normally subtracts directly from the power available for conversion. If corners can be made with radius of curvature larger than or about equal to the stroke of local water particle motion, the viscous losses are usually negligible [1, 11]. The design should be such that this is the case for average waves at the site of operation.

When we talk about the size of buoys in general, it should be understood as their horizontal extension relative to the predominant wavelength λ unless otherwise specified. The following differentiation may be applied:

- Less than $\lambda/6$: small body
- Between $\lambda/6$ and $\lambda/2$: medium-sized body
- Larger than $\lambda/2$: large body

For small buoys the shape does not matter much as long as viscous losses are avoided: In terms of wave radiation pattern, small buoys will behave similar to an axisymmetric body whatever the shape. This is because its wave radiation may be approximated by that of a point source, or pair of point sources. What matters for such buoys is the available volume stroke. In average, the power that can be absorbed will roughly be proportional to the available volume stroke. The size of the buoy should then preferably be chosen large enough to absorb a substantial part of the available power, but small enough to work on full stroke in normal-sized waves. A Budal diagram can be useful in finding a suitable buoy size for a given location, see the Sect. 1.1.7 for further discussion on these volume stroke and the Budal diagram.

Larger buoys that are not axisymmetric become directional, in the sense that the radiated wave field will differ from that of an axisymmetric body (also discussed below). Increased hydrodynamic efficiency can then be achieved by shaping the buoy such as to generate waves along the predominant wave direction. This is the working principle of a terminator device, which is illustrated in Fig. 6.7. As long as the terminator device is made up of a series of units such as paddles, OWC chambers or similar, there is no obvious limit to its useful size.

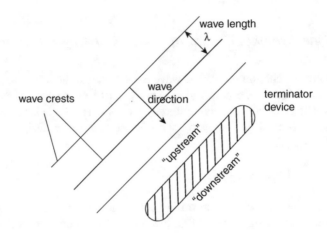

Fig. 6.7 Terminator device. The longest horizontal extension is parallel to the wave crests of waves in the predominant direction

In analogy with the reflecting wall example above (Fig. 6.1): if the body is so large and deep that it reflects most of the incident waves, its motion should induce waves travelling upstream to cancel the reflected waves. On the other side—for bodies that are almost transparent to the incident waves, only waves travelling in the downstream direction need to be generated in order to absorb energy. In practice, we usually have a combination of these two cases.

For axisymmetric bodies, the wave excitation typically increases strongly with width up to an extension of around $\lambda/2$. For bodies beyond this size the increase in size is not paid off by an increase in excitation. This is due to opposing forces over the body surface, making such large bodies less hydrodynamically efficient than smaller bodies. In principle, the same wave radiation pattern as generated by large bodies may be achieved with a number of small bodies placed in a matrix layout, or in a line layout where the each body oscillates in heave and (at least) on more mode of motion.

Whether small or large, the part of the body that is to give the excitation must be found close to the sea surface. Bodies that are placed deep in the water, or which do not have considerable body surface area close to the sea surface, will not be able to absorb much wave energy. This follows directly from how the water moves in a passing wave, with orbital motion (and corresponding dynamic pressure) of decreasing amplitude as we go deeper in the water, cf. Chap. 3 (The Wave Energy Resource) and Fig. 6.8.

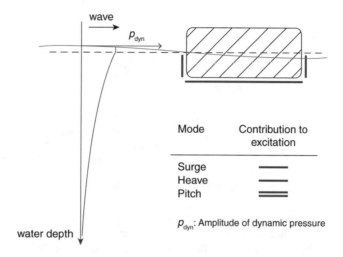

Fig. 6.8 Parts of the body surface area that contributes to surge (*vertical*), heave (*horizontal*) and pitch (both) excitation. The *curve* to the *left* shows the amplitude of hydrodynamic pressure versus distance to mean surface level

6.1.6.2 Heave, Surge or Pitch?

Although it is possible to convert energy also through other modes of motion, heave, surge and pitch are usually the modes considered in practice.

Heaving buoys and bodies that pitch about an axis close to the mean surface level naturally have high hydrostatic stiffness. Recalling the expression for relative bandwidth given above we see that: Unless provided with some means of reducing the stiffness or controlling the motion, such heaving or pitching systems will have quite narrow response bandwidth, which makes them hydrodynamically inefficient wave absorbers in varying irregular seas. This flaw may be mitigated by active use of the machinery through a proper control strategy, or by including mechanical components to counteract the hydrostatic stiffness.

Pitching about an axis close to the surface is less volumetric efficient than surging when it comes to absorbing power [12]. This is due to the fact that such pitch motion gets its excitation from an area distributed along the direction of propagation for the wave, whereas the surge motion gets its excitation mainly from areas of opposing vertical walls a distance apart. This is illustrated in Figs. 6.8 and 6.9. On the other hand, it may be easier to design a practical machinery for pitching bodies than for surging bodies.

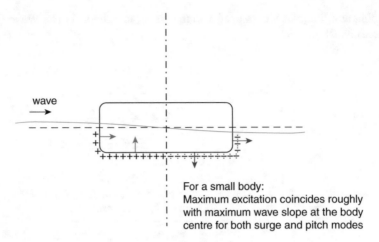

For a small body:
Maximum excitation coincides roughly
with maximum wave slope at the body
centre for both surge and pitch modes

Fig. 6.9 Illustration of the hydrodynamic pressure along a body

For small bodies, heaving motion is the most volumetric efficient. This is because the heave excitation comes from a difference between atmospheric pressure at the top of the body and the full amplitude of hydrodynamic pressure at the bottom, see Fig. 6.8. Surging and pitching take their excitation from a difference in hydrodynamic pressure along the wave, which is quite weak when the body is small. For large bodies the opposite is true, such that pitching and in particular surging motion are favourised over heaving motion.

Systems combining power extraction from two or three modes of motion have the potential of giving a more efficient use of the installed structure [5].

Pitching about an axis close to the bottom is hydrodynamically similar to surging. The surge, sway and yaw modes have no restoring forces, and station-keeping forces must be supplied by moorings or other.

6.1.6.3 Some Examples

Based on the principles explained above, we may think of some examples of systems that would be hydrodynamically good for absorbing wave energy:

- Heaving vertical cylinder of relatively low-draft
- Oscillating volume of small submergence
- Surface-piercing surging flap/bottom-hinged flap
- Large surging bodies.

On the other hand, the following examples should be expected to have poor hydrodynamic preformance:

- Heaving deep-draft cylinders: The volume change takes place too far from the water surface to give considerable wave radiation
- Submerged rigid bodies at considerable depth: Same reason as above
- Large heaving bodies, of diameter larger than about $\lambda/3$. Increasing the diameter above this size would increase costs but only weakly increase the excitation.
- Small pitching bodies
- Submerged surging flap

It should be emphasized that the hydrodynamically best-performing system does not necessarily provide the lowest cost of delivered energy.

6.1.6.4 Comments on Alternative Principles of Power Extraction

The discussion in this chapter has focused on the hydrodynamics of oscillating rigid bodies and also touched upon oscillating water columns. There are other ways of extracting power from ocean waves that may show to be worthwhile. These include:

- **Flexible bodies**, for example in the form of flexible bags (cf. the Lancaster flexible bag device) or tubes (cf. the Anaconda device). These interact with the incident waves through an oscillating volume, and in that sense have features in common with heaving semi-submerged buoys.
- **Overtopping devices**, for example designed as ramps or tapered channels. Their relation to oscillating bodies and the wave interference description is somewhat more obscure, although the overtopping may be seen as a local change of fluid volume that would generate waves to interfere with the incident waves. Hydrodynamically, their wave absorption is usually treated as a wave kinematics problem.
- **Hydrofoils**, or other devices that produce pressure differences over slender members. The principle may typically be to install a system of moving hydrofoils that are relatively transparent to the incident waves, but that is used to radiate a wave field to partially or fully cancel the transmitted waves. Such hydrofoils have been successfully used to propel vessels on wave power [13].

These principles of extraction will not be discussed any further here.

6.1.7 Power Estimates and Limits to the Absorbed Power

The gross power absorbed from the sea can generally be estimated in two ways:

1. By considering the incident and resultant wave fields. The difference in wave energy transport by the two tells how much power has been absorbed.
2. By calculating the product of machinery force and velocity.

This requires a quite elaborate modelling of the wave-absorbing system or extensive experimental campaigns. Simplified and rough estimates of the average absorbed power may be found:

- From experience data based on similar systems. A commonly used measure is the *relative absorption width* or *length*, also referred to as the *capture width ratio*, where experience indicates that values between 0.2 and 0.5 are typical across the wide range of converter designs proposed [14]. It is expected that somewhat higher numbers can be reached with improved conversion systems combined with efficient control algorithms as these are in general both still at a low level of maturity. Average capture width ratios higher than 1.0 have been shown to be realistic for operation of point absorbers in irregular waves [15].
- From theoretical upper bounds on the power that can be absorbed by oscillating bodies. These will be treated in the following.

Firstly, the power is *limited by the radiation pattern* for waves generated by the oscillating system. It is useful to look at this limit for two idealised cases: (i) An *axisymmetric* body, which is symmetric about the vertical axis, and (ii) a large body of width comparable to or larger than the wavelength, often referred to as a *terminator* device. The first type will radiate circular waves when oscillating in heave, and dipole-pattern waves when oscillating in surge or pitch. The second type will radiate plane waves over a limited width, see Fig. 6.10 for illustrations. These properties result in the following limits to the power that can be absorbed [5]:

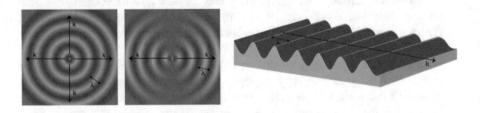

Fig. 6.10 Illustration of radiation patterns: source (*left*), dipole (*middle*) and terminator (*right*) patterns. Reproduced from [16] by courtesy of Elisabet Jansson

(i) Axisymmetric body: $\bar{P} \leq \alpha J \frac{\lambda}{2\pi}$. This expression implies proportionality to wave period cubed and wave height squared. The parameter α is 1 for heave oscillation and 2 for oscillation in surge or pitch.

(ii) Terminator body: $\bar{P} \leq J\, d$. This expression implies proportionality to wave period wave height squared.

Here, J [W/m] is the wave energy transport, λ is the wave length and d is the width of the terminator body or device.

As mentioned, small bodies behave as axisymmetric bodies over the range of wave periods where the size is small compared to the wavelength.

Secondly, there will always be *limits on the available stroke*. These can be caused by the finite volume of the buoys used, or by limits on the stroke of the machinery, often referred to as *amplitude restrictions*. The stroke limitation leads to an upper bound on the average absorbed that is proportional to the wave height H. The bound further depends on the mode of oscillation [12]:

- For heave mode, $\bar{P} < \frac{\pi}{4} \rho g\, V_s \frac{H}{T}$, often referred to as *Budal's upper bound*.
- For surge mode, $\bar{P} < 2\,\pi^3\, \rho\, V_s \frac{H}{T^3} l$
- For pitch mode, $\bar{P} < \frac{2}{3}\pi^3\, \rho\, V_s \frac{H}{T^3} l$

Here V_s is the available volume stroke (illustrated in Fig. 6.11), T is the wave

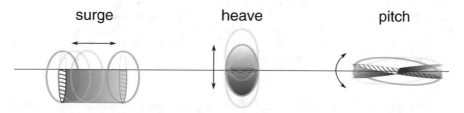

Fig. 6.11 Illustration of the available volume stroke for the different modes of motion. For heave motion, the volume stroke is the body volume itself if not limited by the stroke of the machinery. For surge and pitch motion, the machinery stroke usually sets the limit for the volume stroke

period and l is the length of the device along the direction of wave propagation.

As seen, this second upper bound is inversely proportional to the wave period for heave motion, and to the wave period cubed for surge and pitch motion.

The graphs in Fig. 6.12 illustrate these limitations to the maximum absorbed power. We may refer to these as "Budal diagrams".

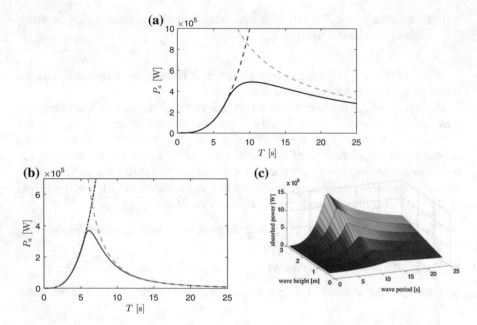

Fig. 6.12 Upper bounds for a heaving semi-submerged sphere (**a**), and for a surging semisubmerged sphere (**b**), both of radius 5 m, and with an incident wave amplitude of 1.0 m. Fully drawn lines shows the absorbed power curves for an optimally controlled buoy. Budal diagram for the heaving sphere extended to also include variation in the wave height H (**c**). All the diagrams are based on a stroke limit of ±3 m

Estimates for the delivered energy from a power plant should always include losses introduced by the power conversion equipment. These will depend strongly on the type of machinery used. It must also be remembered that, in order to approach the limits described above, the motion of the buoys need to be close to optimal. This is the topic of the next section.

6.1.8 Controlled Motion and Maximisation of Output Power

The energy absorbed from a wave by an oscillating body only depends on the motion of the hull relative to the wave. Imagine that you could force the body to move exactly as you wanted. For maximum power extraction, there would be an optimum motion path (position and velocity) that you should try to follow, and that would obviously depend on the incident wave. We may call this the *optimum trajectory* in space and time given the incoming wave.

When designing the wave energy converter system, we should try to make the system such that its response to incoming waves is close to the optimum trajectory. The response is governed by the combination of inertia, stiffness, damping and machinery (PTO) forces. This means that the response may be improved either by the design of the buoy (inertia, stiffness and wave damping), or by using the machinery to get closer to the optimum trajectory. In practice we often use a combination of the two.

Speaking in terms of regular waves, the optimum trajectory for a single-mode absorber may be specified in terms of relative phase and relative amplitude between the wave and the buoy oscillation: The velocity should be in phase with the excitation force, and the amplitude should be adjusted to give the correct inter-ference with the incident wave. The amplitude may be adjusted by changing the damping applied by the machinery.

If in resonance with the incident wave, and with a correctly adjusted machinery loading, a single-mode absorber will automatically obtain the optimum trajectory in the unconstrained case. This is because, at resonance, the optimum phase condition is fulfilled. This does not mean, however, that the buoy needs to be in resonance to perform well. What matters is the phase, or timing, of the motion relative to the incident wave. Systems that are designed to have a large response bandwidth will perform well also off the resonance period. Wave energy converters based on hydro-mechanical systems with low stiffness, such as surging buoys and flaps, will inherently have a large response bandwidth. Others must include mechanical solutions or control strategies to widen the response bandwidth.

References

1. Falnes, J.: A review of wave-energy extraction. Mar. Struct. **20**, 185–201 (2007). doi:10.1016/j.marstruc.2007.09.001
2. Lundby, L. (ed.): Havromsteknologi – et hav av muligheter, NTNU Institutt for Marin teknikk and Fagbokforlaget Vigmostad and Bjørke AS. ISBN: 978-82-321-0441-3
3. Faltinsen, O.M.: Sea Loads on Ships and Offshore Structures. Cambridge University Press (1990)
4. Tupper, E.C.: Introduction to Naval Architecture, 4th edn. Butterworth-Heinemann (2004)
5. Falnes, J.: Ocean Waves and Oscillating Systems: Linear Interactions Including Wave-Energy Extraction. Cambridge University Press, Cambridge (2002). ISBN -13: 978-0521782111
6. Newman, J. N.: Marine Hydrodynamics. MIT Press, Cambridge, Massachusetts (1977). ISBN: 9780262140263
7. Environmental conditions and environmental loads, Recommended practice DNV-RP-C205 (2010), Det Norske Veritas
8. Sundar, V., Moan, T., Hals, J.: Conceptual design of OWC wave energy converters combined with breakwater structures. In: Proceedings of the ASME 2010 29th International Conference on Ocean, Offshore and Arctic Engineering (2010)
9. Evans, D.V.: The oscillating water column device. J. Inst. Math. Appl. **22**, 423–433 (1978)
10. Evans, D.V.: Wave-power absorption by systems of oscillating surface pressure distributions. J. Fluid Mech. **114**, 481–499 (1982)
11. Keulegan, G.H., Carpenter, L.H.: Forces on cylinders and plates in an oscillating fluid. J. Res. Nat. Bur. Stan. **60**, 423–440 (1985)
12. Todalshaug, J.H.: Practical limits to the power that can be captured from ocean waves by oscillating bodies. Int. J. Mar. Energy (2013). doi:10.1016/j.ijome.2013.11.012
13. http://www.wavepropulsion.com/

14. Babarit, A., Hals, J.: On the maximum and actual capture width ratio of wave energy converters. In: Proceedings of the 9th European Wave and Tidal Energy Conference, Southampton, UK (2011)
15. Report on the construction and operation of two Norwegian wave energy power plants, Bølgekraftverk Toftestallen, Prosjektkomiteens sluttrapport, 31.12.1987. Released for public use from the Norwegian ministry of petroleum and energy in 2001
16. Jansson, E.: Multi-buoy Wave Energy Converter: Electrical Power Smoothening from Array Configuration. Master thesis, Uppsala Universitet (2016). ISSN: 1650-8300, UPTEC ES16 009

Chapter 7
Mooring Design for WECs

Lars Bergdahl

7.1 Introduction

7.1.1 General

It would be reasonable that ocean energy devices were designed for the same risk as the platforms in the oil industry. Risk should then be evaluated as a combination of probability of failure and severity of consequences, which means that a larger probability of failure for ocean energy devices would be balanced by the less severe consequences.

The question of some relaxation in safety factors for moorings of WECs has been addressed in the EU Wave Energy Network [1] and at least three times at EWTEC conferences 1995 [2], 2005 [3] and 2013 [4]. Here we will not discuss this but will stick to the present DNV-OS-E301 POSMOOR [5] rules as advised in the Carbon Trust Guidelines [6].

Irregular wave effects are often computed by multiplication of a wave spectrum, for each frequency, with the linear response ratio in that frequency. For instance, using the motion response ratios a response spectrum of the motion will be produced. Thereafter statistical methods can be utilized to assess characteristics of responses in each sea state or in all anticipated sea states during e.g. 50 years.

For large or steep waves and large relative motions non-linear time-domain or non-linear frequency-domain methods must be used, which is out of scope of this chapter.

The goal of the chapter is that the reader shall be able to self-dependently make a first, preliminary analysis of wave-induced horizontal forces, motions and mooring

L. Bergdahl (✉)
Chalmers University of Technology, 41296 Göteborg, Sweden
e-mail: lars.bergdahl@chalmers.se

© The Author(s) 2017
A. Pecher and J.P. Kofoed (eds.), *Handbook of Ocean Wave Energy*,
Ocean Engineering & Oceanography 7, DOI 10.1007/978-3-319-39889-1_7

159

tensions for a moored floating WEC. Necessary prerequisites to attain that goal are the understanding of the physical phenomena, awareness of simplifying assumptions and some insight into the available mathematical and numerical tools.

7.1.2 Mooring Design Development Overview

The development of a mooring system will require different steps:

- Defining the environmental conditions
- Perform a quasi-static analysis, requiring to fine-tune the main mooring design parameters in order to fulfil the design rules. The quasi-static design loop for a mooring system is outlined in Fig. 7.1.

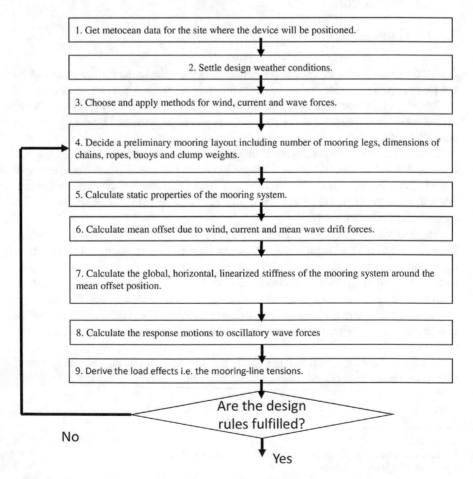

Fig. 7.1 Flow chart of the quasi-static mooring design loop

- Perform model testing to confirm the preliminary quasi-static design. One must then take into account that the moorings may not be correctly modelled due to limitations in water depth or tank width, but some tricks have to be introduced, with springs in the mooring cables compensating for missing cable lengths. Also in some smaller tanks, maybe, the wind and current force cannot be modelled. One then has to preload the mooring system with the calculated mean wind and current force using e.g. a soft horizontal, pre-tensioned spring. Also precaution must be taken concerning the drift force modelling, as small reflexions from the down-wave end of the wave tank may influence the wave drift forces.

  The model testing is based on the assumption that the most important phenomena are governed by potential flow and thus can be modelled using Froude scaling. Drag forces on the whole platform and on the mooring cables are thus not correctly modelled.

- Perform sophisticated dynamic simulations. Such dynamic simulations shall include time varying wind forces, slowly varying wave-drift forces and time series of wave forces. The current could still be considered constant as usually simulations are made for less than three hours duration. Dynamic calculations may first be run in the frequency domain to be able to run many cases. In the end a few critical cases should be run in the time domain.

- Perform prototype tests in the real environment to finally validate the design.

The main design rule is (usually) that the mooring system will be able to ensure the station keeping of the device. In other words, this means that the mooring system will not be overloaded, in terms of tensions in the mooring system and offset of the WEC during the most extreme event it is designed for—usually a 100 year wave.

Comments to the flow chart:

- Weather data may be taken from archived ship born observations, wave buoy data or satellite observations. Wave data can also be "hindcasted" by wave generation models from historical meteorological wind data and also extrapolated by such models to places close to the coast from measurements at off-coast places by a wave generation propagation model like e.g. SWAN [7] or MIKE SW [8]. New measurements may then be started to check the appropriateness of the wave-generation models.

- For the mooring design, usually, combinations of 10-year and 100-year conditions for wind, currents and waves are used, see further Sect. 7.2.4. Before the design conditions are locked it may be wise to confer with the authority or classifying society that will finally verify the mooring design.

- In a quasi-static design mean wind, current and wave drift forces are used for the mean offset and the oscillatory wave force for the dynamic motion response.

7.1.3   Wave-Induced Forces on Structures

One may say that there are two fundamentally different ways to calculate wave-induced forces on structures in the sea. In one method one considers the structure as a whole and assesses the total wave force from empirical or computed coefficients applied on water velocities and accelerations in the undistorted wave motion. This method may be used if the size of the structure is smaller than a quarter of the actual wavelength.

In the other method the pressure distribution around the surface of the structure is computed taking into account the effect on the water motion distorted by the structure itself, and subsequently integrated around the structure. Both these approaches are used for the oscillating wave forces in Sect. 7.3.3.2

In both cases some mathematical model for describing the wave properties is necessary. For instance, by making the simplified assumption that the wave motion can be regarded as potential flow, velocities, accelerations and water motion can be computed in any point under a gravity surface wave by a scalar quantity, the velocity potential.

7.1.4   Motions of a Moored Device in Waves

A moored device in waves will be offset by steady current, wind and wave drift and will oscillate in six degrees of freedom. In very long waves its motion will just follow the sea surface motion with some static reaction from the mooring system, but for shorter waves—near the horizontal and vertical resonances of the body-mooring system—the motion may be strongly amplified and out of phase with the sea surface motion. For still shorter waves the motions will be opposed to the wave motion but less amplified, so when the crest of the wave passes the device the device will be at its lowest position, with obvious consequences for water overtopping the device, or air penetrating under the bottom of the device. For very short waves the wave forces will be completely balanced by the inertia of the device itself and will show negligible motion. Methods for estimating motions of floating objects are described quantitatively in Chap. 10.

7.2   Metocean Conditions

7.2.1   Combinations of Environmental Conditions

The target probabilities of failure and return periods for extreme forces as given in DNV-OS-E301 [5] (POSMOOR) are referred in Tables 7.1 and 7.2. These will be used here as approved, although it may seem reasonable that the safety and

Table 7.1 Target annual probability of failure. For consequence-class definitions see Sect. 7.5.1.2

Limit state	Consequence class	Target annual probability of failure
ULS	1	10^{-4}
ULS	2	10^{-5}

Table 7.2 Return periods for environmental conditions

Return period		
Current	Wind	Waves
10	100	100

reliability requirements for offshore hydrocarbon units exceed those that should be applied to floating ocean wave energy converters.

7.2.2   Design Wave Conditions

According to DNV-OS-E301 [5], sea states with return periods of 100 years shall normally be used. The wave conditions shall include a set of combinations of significant wave height and peak period along the 100-year contour. The joint probability distribution of significant wave height and peak wave periods at the mooring system site is necessary to establish the contour line. If this joint distribution is not available, then the range of combinations may be based on a contour line for the North Atlantic. It is important to perform calculations for several sea states along the 100-year contour line to make sure that the mooring system is properly designed. For instance, moored ship-shaped units are sensitive to slowly varying, low-frequency wave forcing. Therefore, in sea states with shorter peak periods, 6–10 s, the slowly-varying drift force may excite large resonant surge motions, while in a sea state with a long peak period around 20 s the motion is dominated by the wave-frequency motion and the overall damping is larger preventing resonant motion. How to choose sea states along the contour line is indicated in Fig. 7.2. The same values for wind and current shall be applied together with all the sea states chosen along the 100-year contour.

If it is not possible to develop a contour line due to limited environmental data for a location a sensitivity analysis with respect to the peak period for the 100 year sea state shall be carried out. The range of wave steepness criteria defined in DNV-RP-C205 [9] (Paragraph 3.5.5) can then be applied to indicate a suitable range of peak wave periods to be considered in the sensitivity analysis. The JONSWAP spectrum is a reasonable spectrum model for

$$3.6 < T_p/\sqrt{H_s} < 5, \tag{7.1}$$

Example of characteristic environmental conditions for rule check

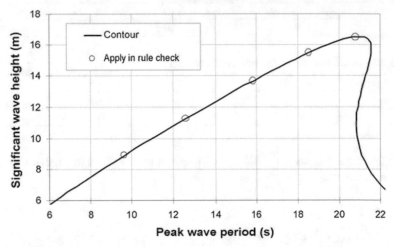

Fig. 7.2 Selections of sea states along a 100-year contour line. (DNV-OS-E301 [5])

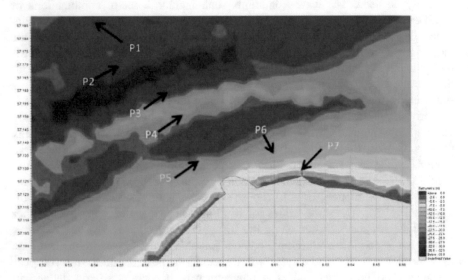

Fig. 7.3 Location of the 7 points with wave data at DanWEC

but should be used with caution outside this range. In the guidance notes in POSMOOR some 100 year contour lines for offshore sites are given. However, they are not very useful in wave energy contexts as wave-energy sites are closer to the coast in shallower areas with milder wave climates. Therefore, it is mostly necessary to use site-specific data, which can be created by using offshore data and a spectral wave model as SWAN [7] or MIKE 21 SW [8] for transferring the deep water statistics to specific near-shore sites.

7.2.3   Environmental Data at DanWEC

Within the SDWED project, DHI produced data for Hanstholm [10], using MIKE 21 SW [Mean wind speed is taken]. This data will be used for the example mooring design.

The wave conditions for seven points off Hanstholm, Fig. 7.3, have been calculated from the DHI-data by Pecher and Kofoed [11] and are referred in Table 7.3. The individual maximum 100 year wave ($1.86H_s$) may be depth limited as conventionally is approximated by $H_{max} < 0.78h_d$, but at the intended site for the example WEC buoy the water depth is 30 m, why the waves at this site are not depth limited.

Table 7.3 Waves at DanWEC, in front of Hanstholm

		Average Wave conditions				Design wave, Hm0(m)			
	Water depth	Hm0	Tp	T02	Pwave	Return period* (years)			
Location	(m)	(m)	(s)	(s)	(kW/m)	100	50	20	10
P1	29	1.25	6.4	4.2	9.4	9.5	9.1	8.5	8.0
P2	27.5	1.23	6.4	4.2	8.9	9.3	8.9	8.3	7.8
P3	32	1.19	6.3	4.1	8.1	8.8	8.4	7.8	7.4
P4	18.5	1.18	6.4	4.2	8.3	8.9	8.5	7.9	7.4
P5	19	1.09	6.3	4.1	6.8	7.8	7.5	7.1	6.7
P6	14	0.97	6.4	4.0	5.4	6.0	5.9	5.8	5.6
P7	5	0.74	6.6	4.4	2.9	2.7	2.6	2.6	2.5

We also need the design wind, water level and current conditions. Wind and water levels are reported by Sterndorf in a report for WavePlane, and are given below in Tables 7.4 and 7.5.

Table 7.4 Design wind conditions [12]

Probability of exceedance	Wind speed for wind coming from				
	SW	W	NW	N	NE
$V_{wind,3\,h}$ – 1 year (m/s)	21.0	25.0	25.0	19.0	20.0
$V_{wind,3\,h}$ – 10 year (m/s)	24.0	30.0	29.5	23.5	25.0
$V_{wind,3\,h}$ – 100 year (m/s)	28.0	34.0	33.0	28.0	29.0
Probability of wind direction (%)	15.5	18.4	11.8	5.2	8.4

Table 7.5 Design water
levels [12]

Probability of exceedance	High water (m)	Low water (m)
3 h 1 year	1.22	1.28
3 h 10 year	1.58	1.52
3 h 100 year	1.96	1.78

Sterndorf gives the wind speed as $V_{wind,\ 3\ h}$, but normally the 10 min mean value is used for mooring design of floating objects.

Sterndorf [12] estimates the current to 3 % of the wind speed, assuming the current to be locally wind generated, yielding 0.68 m/s from SW and 0.58 m/s from NE, while Margheritini [13] cites measured values at 0.5–1.5 m/s coast parallel.

7.2.4 Example Design Conditions

In the sample design calculations below the following values are chosen:

- Mean wind speed is taken from Table 7.4, 100 year return period: $U_{10\ min,10\ m} = 33$ m/s.
- Mean current velocity is set to the maximum measured value according to Margheritini. See text below Table 7.5: 10 year return period: $U_c = 1.5$ m/s
- Waves are taken with guidance from Table 7.3 as representative of Point 3, 4 and 5 to: 100 year return period: $H_s = 8.3$ m.
- A PM-type spectrum as a Bredtscneider or an ISSC-spectrum then gives $T_p = 12.9$ s and $T_{02} = 9.2$ s $< T_z < T_{01} = 9.9$ s. The probable maximum wave height of 1000 waves is then around $H_{max} = H_s 1.86 = 15.4$m.
- Wind, current and waves are acting in the same direction.
- Water depth is taken as $h_d = 30$ m from Pecher et al. [14]

7.3 Estimation of Environmental Forces

7.3.1 Overview and Example Floater Properties

It is demanding to establish the hydrodynamic forces for WECs, because they may undergo very large resonant motion, have very complex shapes composed of articulated connected bodies or involve a net flow of water through the device. This makes it difficult to use conventional potential methods. Probably, most devices need to undergo extensive tank and field testing. However, here we will sketch simplified methods for first estimates of forces useful in the concept stage and for planning tank tests.

In order to design a mooring solution, all environmental forces need be included that can have a significant influence on the motions of the floating body and thereby on the mooring response. The main ones are:

- Wind force
- Sea current force
- Wave forces: Both mean wave drift forces and oscillatory wave forces

The following paragraphs will introduce how these can be estimated for a floating, moored, vertical, truncated, circular cylinder with properties according to Table 7.6.

Table 7.6 Properties of the sample floater

Diameter	(m)	5
Height above mean water surface	(m)	5
Draught	(m)	5
Mass	(tonne)	100
Pitch inertia around mean water surface	(tonne m^2)	1830
Cross coupled inertia ($m_{24} = m_{42} = -m_{15} = -m_{51}$)	(tonne m)	243

7.3.2 Mean Wind and Current Forces

7.3.2.1 Introduction

According to DNV-OS-E301 [5] the wind and current force should be determined by using wind tunnel tests. Wind forces from model basin tests are only applicable for calibration of an analysis model, while the current forces may be estimated from model basin tests or calculations according to recognised theories (DNV-RP-C205 [9], Sect. 7.6.6). In preliminary design also wind forces calculated according to recognised standards may be accepted, such as in DNV-RP-C205 [9], Sect. 7.6.5.

The mean wind and drag force may be calculated using a drag force formulation, with drag coefficients from model tests, or numerical flow analysis. Wind profile according to DNV-RP-C205 [9] and ISO19901-1 shall be applied. Oscillatory wind forces due to wind gusts shall be included:

$$F = CA\frac{1}{2}\rho U^2 \tag{7.2}$$

Here C is traditionally called the shape coefficient for wind force calculations and drag coefficient for current force calculations, A is the cross sectional area projected transverse the flow direction, ρ is the density of the fluid and U is the fluid velocity at the height of the centre of the exposed body. Here we will use the design 10 min mean for the air velocity and the design value of the current, as the response of the horizontal motions and the induced mooring tension are in this time scale.

Values on the coefficient C for different shapes are given in DNV-RP-C205 [9], but can also be found in other standard literature like Faltinsen [15], Sachs [16]. For more complicated superstructures a discussion is found in Haddara and Guedes-Soares [17]. In DNV-RP-C205 there are also guidelines for calculating vibrations or slowly varying wind force due to a wind spectrum.

Below the calculation of the wind and current forces are sketched but more detailed information can be found in DNV-RP-C205.

7.3.2.2 Wind Force on the Sample Floater

Mean wind speed $U_{10\,\text{min},10\,\text{m}} = 33$ m/s.

To use the drag force expression Eq. 7.21 for the wind force we must first estimate the wind speed at the centre of the buoy which is situated 2.5 m above the mean water surface. The wind is given at 10 m height. A wind gradient expression giving the wind speed at 2.5 m from the value at 10 m gives:

$$U(2.5\,\text{m}) = U(10\,\text{m})\left(\frac{2.5\,\text{m}}{10\,\text{m}}\right)^{0.12} = U(10\,\text{m})0.85 = 28.9\frac{\text{m}}{\text{s}} \qquad (7.3)$$

In order to estimate the shape coefficient C from graphs and tables in DNV-RP-C205 we must also calculate the Reynolds number:

$$Re = \frac{U_{T,z}D}{v_a} = 9.6 \times 10^6 \qquad (7.4)$$

where $D = 5$ m is the diameter and v_a is the kinematic viscosity $= 1.45 \times 10^{-5}$ m^2/s (DNV-RP-C205 [9], APPENDIX F)

Figure 7.25 in DNV-RP-C205 gives C = 1.1 for a relative roughness of 0.01.

The aspect ratio is $2h_b/D = 2$ and gives a reduction factor of $\kappa = 0.8$ for supercritical flow. The height above the water surface of the buoy, h_b, is the same as the diameter, D, and it is considered as mirrored in the water surface to calculate the aspect ratio, which is defined as the length over width ratio.

Thus the wind force is (Air density $\rho_a = 1.226$ kg/m^3 at 15 °C)

$$F_a = \kappa CDh_b\frac{1}{2}\rho_aU_{T,z}^2 = 10.5 \text{ kN} \qquad (7.5)$$

7.3.2.3 Current Force on the Sample Floater

The current speed is assumed to have no vertical gradient close to the free water surface and the mean current speed $U_c = 1.5$ m/s. In order to estimate the drag coefficient C from graphs and tables in DNV-RP-C205 we must estimate the

Reynolds number. Diameter $D = 5$ m, and the kinematic viscosity $v_w = 1.19 \times 10^{-6}$ m^2/s, thus the Reynolds number is $Re = \frac{U_c D}{v_w} = 6.3 \times 10^6$.

Again Fig. 7.25 in the DNV-RP-C205 gives again C = 1.1 for a relative roughness of 0.01

The aspect ratio is $2D_b/D = 2$ and gives a reduction factor of $\kappa = 0.8$ for supercritical flow. The draught below the water surface of the buoy, D_b, is the same as the diameter, D, and again it is considered as mirrored in the water surface to calculate the aspect ratio.

Thus the current force is (Sea water density $\rho = 1025.9$ kg/m^3 at 15 °C)

$$F_c = \kappa C D D_b \frac{1}{2} \rho U_c^2 = 24.5 \text{ kN} \qquad (7.6)$$

7.3.3   Wave Forces

7.3.3.1   Mean Wave Drift Force

Mean Wave Drift Force in Regular Waves, Simplified Approach

Basically there are two alternative approaches to estimate the wave drift force. The first approach involves integrating the pressure over the instantaneously wetted surface of the body. This will, for a body in a regular wave, give a force composed by a mean force, a force at the same frequency as the incident wave (the usual first-order wave force, which will be discussed in the next section) and a force at the double frequency. For the slowly varying drift forces only the mean force is of interest. The second approach involves utilising the momentum conservation and will be used here. We will approximate it in 2D for a terminator type body subjected to a plane, unidirectional wave motion with the incident wave amplitude a.

Through a vertical the time mean of the incident momentum per unit width of structure is

$$I_0 = \frac{\rho g a^2}{4} \qquad (7.7)$$

If this wave is blocked by a vertical wall, a wave with the same amplitude, $r = a$, will be reflected in the opposite direction and the momentum acting on the wall, or mean drift force will become

$$F_d = I_{in} - I_{out} = \frac{\rho g}{4}(a^2 + r^2) = 2I_o = \frac{\rho g a^2}{2} \qquad (7.8)$$

This is the largest possible mean wave drift force on a floating body per unit width of structure. For a floating 2D body, however, not all the energy will be reflected and the body will be set in motion and radiate energy up-wave and down-wave. If we denote the amplitude of the combined reflected and back-radiated wave by r and the amplitude of the combined transmitted and down-wave radiated wave by t, then a momentum approach will give

$$F_d = \frac{\rho g}{4}(a^2 + r^2 - t^2)$$ (7.9)

This was set up by Longuet-Higgins [18]. Maruo [19] stated that if there are no losses in the flow, the sum of the powers in the r wave and the t wave must equal the power in the incident wave, i.e. $(a^2 = r^2 + t^2)$ and consequently

$$F_d = \frac{\rho g}{2}r^2$$ (7.10)

For successful WECs this equation is not valid, as then $a^2 \gg r^2 + t^2$ and thus for complete wave absorption in the limit $F_d = \rho g a^2/4$. For a device in standby again Eq. 7.9 is valid.

For real devices with limited transverse extension the above equations can be seen as upper bounds as the wave is scattered around the object and waves are radiated by the object in the horizontal plane.

Mean Wave Drift Force in Irregular Waves

A very simple approach on the conservative side is based on the assumption that the object reflects all waves in the opposite direction to the incoming waves for all component waves, with the amplitude, a_i. In e.g. a PM-spectrum with $H_s = 8.3$ m the drift force would be:

$$F_d = \frac{1}{2}\rho g \sum_i \frac{1}{2}a_i^2 D = \frac{\rho g H_s^2}{32}D = 108 \text{ kN}$$ (7.11)

The above equation presumes that all components would be reflected without any scatter. However an object with a diameter less than one quarter of a wavelength diffracts or reflects negligible energy. In our case this wave length is 4 $D = 20$ m corresponding to the wave period 3 s and frequency 0.28 Hz. Plotting a PM-spectrum with $H_s = 8.3$ m and drawing the line for $f = 0.28$ Hz gives the following picture that indicates that the wave drift force would be negligible, as almost the entire spectrum is below this frequency (Fig. 7.4).

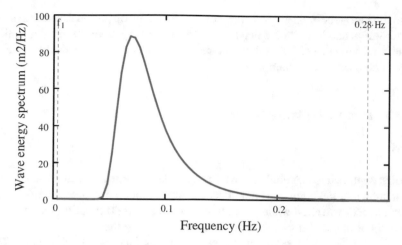

Fig. 7.4 The design wave energy spectrum, PM-spectrum with $H_s = 8.3$ m. The wave period 0.28 Hz corresponding to a deep-water wave length of 4 D is marked in the figure

To check that the drift force really is small in the survival design storm with $H_s = 8.3$ m, we have calculated the drift force coefficient for the floating buoy with WADAM [20] and integrated the total drift force in that sea state, see Fig. 7.5. Using WADAM's definition of the drift-force coefficient, the drift force can be written

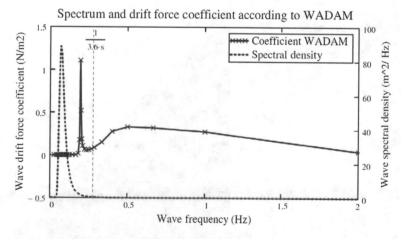

Fig. 7.5 The drift force coefficient as a function of wave frequency as calculated by WADAM [9]. Note the effect of the vertical resonant motion at 0.2 Hz

$$F_d = 2\rho g D \sum_i C_{d_i} \frac{1}{2} a_i^2 \qquad (7.12)$$

The resultant drift force was found to be $F_d = 2.5$ kN, which in this case is 25 % of the estimated wind force and 10 % of the current force and can thus—as a first approximation—be neglected in the design storm. In operational sea states with shorter waves and lower wave heights the drift force may be of the same magnitude as the wind and current forces, but all three forces are smaller. The drift force of 2.5 kN will be used in the example below.

7.3.3.2   First-Order Wave Forces

Overview

The first approach to calculating wave forces on bodies in water was founded on the assumption that the body does not affect the water motion and pressure distribution in the incident wave. Nowadays one would normally use diffraction theory, taking into account the scatter of the incident wave caused by the body.

In Fig. 7.6 we can note different flow regimes as function of $\pi D/\lambda$ and H/D. In the present case $\pi D/\lambda = \pi D/(g\,T_p^2/2\pi) \approx 0.06$ and $H_{max}/D \approx 3$, which set us in the inertia and drag regime. For such bodies with a characteristic diameter of less than 1/4 to 1/5 of a wave length the effect on the wave is small, and the wave force can, as an approximation, be set to the sum of an inertia term and a drag term. The

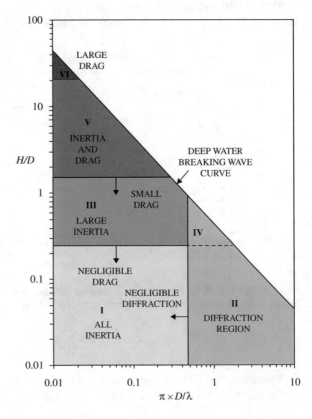

Fig. 7.6 Different wave force regimes (Chakrabarti 1987, cited by DNV).
D = characteristic dimension, H = sinusoidal wave height, λ = wave length. Adapted from DNV-RP-C205 [9]

inertia term is the product of the displaced mass, added mass included, and the undisturbed relative water acceleration in the centre of displacement. The drag term depends on the relative velocity between water and body. In surge this so called Morison formulation is:

$$F = \rho V \frac{du}{dt} - m\ddot{x} + C_m \rho V \left(\frac{du}{dt} - \ddot{x} \right) + \frac{1}{2} C_D \rho A |u - \dot{x}|(u - \dot{x}) \tag{7.13}$$

Where:

- F is the reaction force from e.g. a mooring system (Unmoored body $F = 0$)
- ρ is the density of water
- V the displaced volume
- u and $\frac{du}{dt}$ the undisturbed horizontal water velocity and acceleration in the centre of the body
- m the mass of the body
- x the horizontal position of the body
- $\ddot{x}$ and $\dot{x}$ the acceleration and velocity of the body
- C_m an added mass coefficient (Can be taken from standard values in e.g. DNV-RP-C205 [9])
- C_D a drag coefficient (Can be chosen from recommendations in e.g. DNV-RP-C205)
- A the cross-sectional area in the direction perpendicular to the relative velocity

So far we have not defined any properties of the mooring system, but for the time being we can assume that the body is fixed to select the coefficients C_m and C_D, again using DNV-RP-C205 [9]. One should then take into account the variation of C_D and C_m as functions of the Reynolds number, the Keulegan-Carpenter number and the relative roughness.

$$\text{Reynolds number}: Re = u_{max} D/v$$
$$\text{Keulegan-Carpenter number}: KC = u_{max} T/D$$
$$\text{Relative roughness}: k/D$$

where:

- D = diameter = 5 m
- T = wave period = T_p = 12.9 s
- k = roughness height = 0.005 m
- u_{max} = maximum water velocity in a period $\pi H_{max}/T_p$ = 3.8 m/s (assuming circular water motion in deep water) and
- $v_w = 1.19 \times 10^{-6}$ m²/s = fluid kinematic viscosity.

For the buoy $Re = 8 \times 10^6$, $KC = 10$ and $k/D = 10^{-3}$. For coefficients of slender structures DNV-RP-C205 still refers to Sarpkaya and Isacson [21] but the problem is that their graphs and experience are limited to $Re < 15 \times 10^5$, see also

Chakrabarti [22]. Anyway, these graphs and also equations in DNV-RP-C205, Sect. 7.6.7, point to $C_D = 1$ and $C_m = 1$ for circular cylinders. As for the steady flow the drag coefficient may be reduced to 0.8 due to the aspect ratio. In Appendix D, RP-C205, Table D-2 there is also an indication that C_m could be reduced to around 0.8 due to the aspect ratio $L/D = 2$

Wave Forces on "Small" Bodies D < L/4

Wave Forces in a Regular Wave (Small Body)

Applying the Morison equation above for the fixed body, it reduces to

$$F = \rho V(1 + C_m)\frac{du}{dt} + \frac{1}{2}C_D\rho A|u|u \qquad (7.14)$$

This force, as a function of time for the wave amplitude $a = H_{max}/2$ and period $T = T_p$, is drawn in the figure below together with the horizontal water acceleration. One can note that the evolution in time is affected by the drag, but that the maximum value is almost unaffected, and can approximately be calculated as the mass (inertia) force amplitude:

$$F_M = \rho V(1 + C_m)\frac{du_a}{dt}\Big|_{max} = 0.44 \text{ MN} \qquad (7.15)$$

The mass force amplitudes $F_M = \pm 0.44$ MN are drawn as horizontal lines in the graph. The drag-force maximum is $F_D = 0.3$ MN but is 90 degrees out of phase with the water acceleration and in phase with the water velocity (Fig. 7.7).

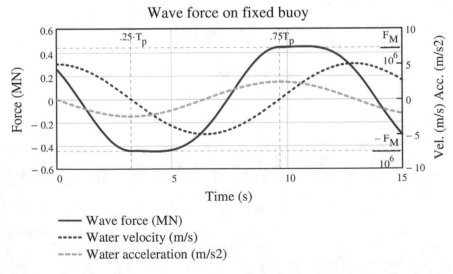

Fig. 7.7 The Morison force as a function of time for wave amplitude $a = H_{max}/2$ and period $T_p = 12.9$ s. The water acceleration is drawn for comparison

We can note that the wave force amplitude is one order of magnitude larger than the mean force from wind, current and wave drift. However, for a floating moored body the wave force would be carried by the inertia of the body and not by the mooring or positioning system as we do not want to counteract the wave-induced motion only prevent the buoy from drifting off its position.

Wave Forces in Irregular Waves (Small Body)

If we neglect the drag term in the wave force equation above, we can calculate the wave force spectrum, $S_F(f)$, directly by multiplication of the wave spectrum, $S_{PM}(f)$ by the square of the wave force ratio, $f_w(f)$. The problem is that for $f > 0.28$ Hz the diffraction would be important and the small body assumption is not valid. The force amplitude divided by the wave amplitude or force amplitude ratio (also known as RAO) would become

$$f_w(f) = \frac{F}{a} = \frac{\rho V}{a}(1 + C_m)\frac{du}{dt}_{max} = \rho V(1 + C_m)gk\frac{\cosh(k(z+h))}{\cosh(kh)} \quad f < 0.28 \text{ Hz and}$$
$$f_w(f) = 0 \qquad\qquad\qquad\qquad\qquad\qquad\qquad\qquad\qquad\qquad f > 0.28 \text{ Hz.}$$
$$(7.16)$$

where $k = 2\pi/L$ is the wave number. In deep water $k = g/\omega^2 = g/(2\pi)^2$.

The wave force spectrum could then be calculated as

$$S_F(f) = (f_w(f))^2 S_{PM}(f) \qquad\qquad (7.17)$$

These functions are drawn in Fig. 7.8

Fig. 7.8 Wave energy spectrum, $S_{PM}(f)$, force amplitude ratio, $f_w(f)$, and force spectrum, $S_F(f)$. Morison approach

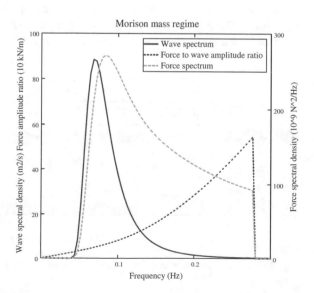

The significant force amplitude is then

$$F_{Msamp} = 2\sqrt{m_{0F}} = 2\sqrt{\int_{0\,Hz}^{0.28\,Hz} S_F(f)df} = 0.38\ MN \qquad (18)$$

And the maximum force in 3 h would be $F_{Mmax} = 1.86\ F_{M\,samp} = 0.71\ MN$.

This is similar to calculating the significant wave height and relation between H_{max} and H_s, but here it is the Rayleigh distribution for the for the amplitudes, that is why there is $2\sqrt{m0}$ and not $4\sqrt{m0}$ in Eq. 7.20.

Wave Forces on "Large" Bodies

Overview

To extend the force calculation to shorter waves or relatively larger bodies we are forced to use diffraction theory, which is more demanding and, yet, does not take drag (viscous) forces into account. On the other hand radiation damping caused by waves generated by the motion of the body in or close to the free surface is included, which lacks in the Morison approach. For the diffraction problem of the vertical circular buoy there are analytical series solutions available e.g. in Yeung [23] and Johansson [24] (Figs. 7.9 and 7.10). Here we will illustrate it by using results from Johansson. Bodies with general form can be calculated in panel diffraction programs like WAMIT [20].

In Figs. 7.9, 7.10 and 7.11, graphs with added mass, radiation damping and wave force amplitude ratio as functions of frequency are displayed. The wave force amplitude ratio will be used immediately for comparison of wave forces on the fixed body. The added mass and radiation damping will be used later for calculating wave motion and slowly varying wave drift motion of the moored buoy (Fig. 7.11).

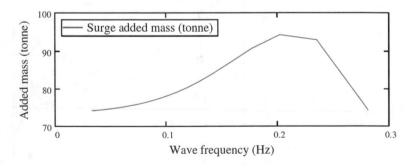

Fig. 7.9 Surge added mass, A_{11}, as a function of wave frequency

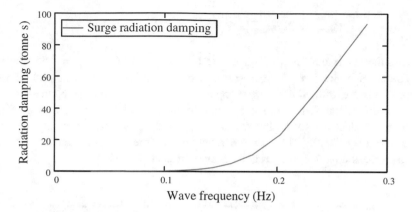

Fig. 7.10 Surge radiation damping, B_{11}, as a function of wave frequency

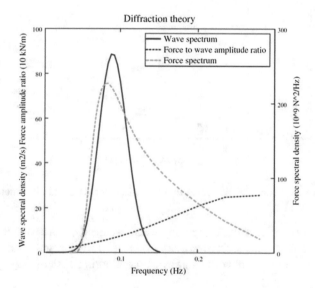

Fig. 7.11 Wave energy spectrum, $S_{PM}(f)$, force amplitude ratio, $f_{dw}(f)$, and force spectrum, $S_{dF}(f)$. Diffraction results from Johansson [24]

Wave Forces in Irregular Waves (Large Body)

The wave force spectrum can now be calculated as before but with diffraction results instead of approximate coefficients

$$S_{dF}(f) = (f_{dw}(f))^2 S_{PM}(f) \tag{7.19}$$

The significant force amplitude is now estimated as

$$F_{dsamp} = 2\sqrt{m_{0dF}} = 2\sqrt{\sum_i S_{dF}(f_i)\Delta f_i} = 0.30 \text{ MN} \qquad (7.20)$$

And the maximum force in 3 h would be $F_{dmax} = 1.86\,F_{dsamp} = 0.55$ MN.

The 23 % reduction of the force is due to the lower force amplitude ratio according to the diffraction theory compared to the Morison model. Note especially that the diffraction force ratio has a maximum around 0.3 Hz in this case and actually will decrease for higher frequencies while the Morison counterpart grows to infinity (Fig. 7.12). This is more realistic than the overestimated force in the Morison mass approach for irregular waves in Sect. 7.3.3.2.2.2

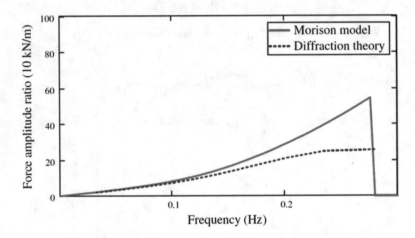

Fig. 7.12 Force amplitude ratio according to the Morison approach and diffraction theory

In the quasi-static mooring design approach, we need estimate the motion of the moored object in regular design waves or in an irregular sea state. To get the mooring force we must know the statics of the mooring system.

7.3.4   Summary of Environmental Forces on Buoy

In Table 7.7 there is a summary of results from the gradually more sophisticated calculations. First one can note that—in this case—the simplest wave-drift estimate gives 40 times as large value as the one founded on diffraction theory. This is important in relation to the wind and current force. The Morison wave force for a regular sinusoidal wave is very dependent on the assumed wave period, while the Morison approach for irregular waves gives some better significance, however some 20 % overestimation.

Shaded values will be used in the design as they are considered as most realistic.

Table 7.7 Key results from force estimates on the floating buoy

Mean loads		Force (kN)	Wave force	Force (MN)	
Wind	33 m/s	10.5	Morison Regul. $H_{max}/2 = 7.7$ m	0.44	Amplitude
Current	1.5 m/s	24.5	Morison mass regime	0.38	Significant
Wavedrift $H_s = 8.2$ m	Simple	108	Irregul. $H_s = 8.2$ m	0.71	Most prob. maximum
	Diffraction	2.5	Diffraction Irregul. $H_s = 8.2$ m	0.30	Significant
Total mean	Simple	143		0.55	Most prob. maximum
	Diffraction	37.8			

Shaded values will be used in the design

7.4   Mooring System Static Properties

7.4.1   Example

For illustrative purposes a mooring configurations will be used as presented by Pecher et al. [14]: a three-leg Catenary Anchor Leg Mooring system, CALM, see Fig. 7.13.

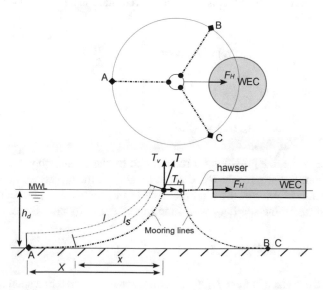

Fig. 7.13 Sketch of a three-leg Catenary Anchor Leg Mooring (CALM). [14]

The CALM system is composed of three chain-mooring legs directly fastened to the example buoy. This is different to the example by Pecher et al. [14] who have assumed that the mooring legs are connected to a mooring buoy, which in turn is coupled by a hawser to a wave-energy device. The legs have equal properties listed in Table 7.8. The lengths of the mooring lines should be chosen such that they will just lift all the way to the anchor when loaded to their breaking load.

Table 7.8 Example properties of the CALM system

Three-leg system 120 deg	Chain Steel grade Q3	Notation
Water depth	30 m	h_d
Horizontal pretension	20 kN	H_p
Unstretched length	509 m	s
Breaking load	2014 kN	T_B
Diameter	50.4 mm	
Mass per unit unstretched length	53.65 kg/m	q_o
Weight in sea water per unit unstretched length	457 N/m	γ_r
Axial stiffness	228 MN	$K = EA$

7.4.2 Catenary Equations

Here we will use the equations for an elastic catenary expressed in the unstretched cable coordinate from its lowest point, or from the touch-down point at the sea bottom as in Fig. 7.13, to a material point, s_o [25].

The horizontal stretched span or the horizontal distance, $x_{o1}(s_o)$, from the touch-down point, $s_o = 0$, is

$$x_{o1}(s_o) = a \arcsinh\left(\frac{s_o}{a}\right) + \frac{\gamma_r a}{K} s_o, \qquad (7.21)$$

and the vertical span is

$$x_{o2}(s_o) = \sqrt{a^2 + s_o^2} + \frac{\gamma_r}{2K} s_o^2 - a, \qquad (7.22)$$

where $a = H/\gamma_r$, i.e. the horizontal force divided by the un-stretched weight per unit length in water. Solving for the lifted cable length, s_o, for $x_{o2}(s_o) = h_d =$ the water depth, we can now express the total distance to the anchor $X(H)$ including the part of chain resting on the sea floor as a function of the horizontal force, H.

$$X(H) = x_{o1}(s_o(H)) + (s - s_o(H))(1 + \frac{H}{K}) \qquad (7.23)$$

or inversely the horizontal force $H(X)$ as a function of the stretched span X, Fig. 7.14

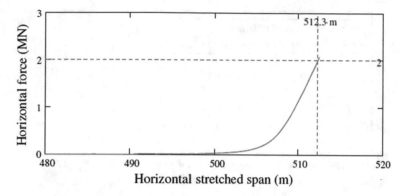

Fig. 7.14 The horizontal force as function of the horizontal, stretched span

In the intended system we have assumed a pretension of $H_p = 20$ kN at zero excursion. This corresponds to a horizontal span of $X(H_p) = 498.36$ m. Finally we can add the reaction of the three legs to get the total horizontal mooring force as a function of the excursion, $x = X(H) - X(H_p)$, in the x-direction in parallel to the upwind leg.

$$F_{tot}(x) = H(x) - 2\cos(60°)H(-\frac{x}{\cos(60°)}) \qquad (7.24)$$

In the example we can see that almost all the horizontal force is carried by the cable in the up-wave direction as soon as the excursion exceeds 4 m.

Last we need calculate the horizontal stiffness, $S(x)$, of the mooring system, that is, the slope of the blue function displayed in Figs. 7.15 and 7.16.

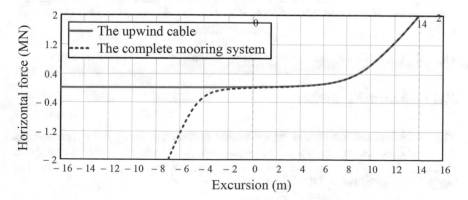

Fig. 7.15 Horizontal force as a function of the excursion of the buoy. The up-wave cable takes most of the force

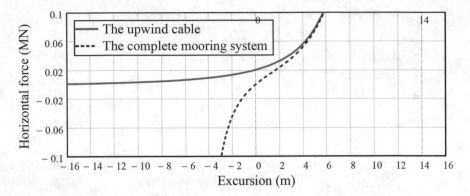

Fig. 7.16 Horizontal force as a function of the excursion of the buoy. Different range of vertical axis compared to Fig. 7.15

It is interesting to note that the stiffness for negative excursion is larger than for positive excursion, which is caused by having two interacting legs in this direction (Fig. 7.17).

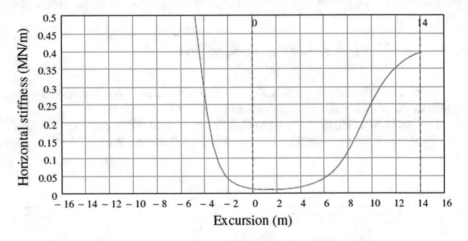

Fig. 7.17 The horizontal stiffness of the mooring system as a function of the excursion

7.4.3 Mean Excursion

The horizontal motion should be calculated around the mean offset (excursion). Therefore the offset due to the mean forces is calculated using the methods described above. We also need the mooring stiffness around the mean offset. The result is given in Table 7.9.

Table 7.9 Summary of offset and mooring stiffness due to the mean environmental forces

Mean force	Force	Mean offset	Tangential Stiffness
	(kN)	(m)	(kN/m)
Wind + current + WADAM wave drift	10.5 + 24.5 + 2.5 = 37.5	2.6	12

7.5 Alternative Design Procedures

7.5.1 Quasi-Static Design

7.5.1.1 Quasi-Static Design Procedure

The most used method for designing mooring systems is still a variant of the quasi-static design procedure, described for instance by Selmer [26] (see Fig. 7.18).

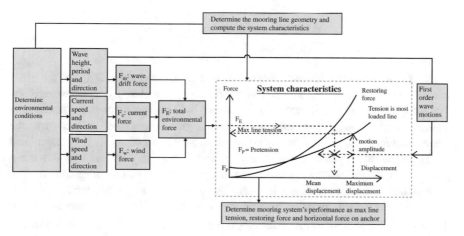

Fig. 7.18 Quasi-static analysis. Adapted from [26]

1. Wind, current and wave-drift forces are considered constant and acting in the same direction.
2. The horizontal reaction force as a function of offset is calculated for the mooring system, and from this the offset and cable tensions due to the constant forces.
3. The motion of the freely floating platform is calculated for the design sea state.
4. The maximum horizontal offset due to the wave induced motions is added to the constant offset, and the corresponding (static) cable tensions are obtained from the static functions calculated in step 2.
5. The tension force in the most loaded cable is compared with the allowed force for operational or survival conditions.

In modern quasi-static procedures, first, constant forces from mean wind, mean current and mean wave drift are assumed acting co-linearly on the moored floating object, as is stated in DNV-OS-E301 POSMOOR [5] of Det Norske Veritas (DNV). This gives a mean horizontal offset in the force direction. The equation of motion for the moored floating object—including the stiffness of the mooring system—is then solved so that possible resonance effects are taken into account. In the original approach, described above, the wave-induced motion for the freely floating platform was used, assuming that the mooring system did not have any effect on the motion. This is not recommended nowadays, but gives small errors for large floating platforms reasonably deep water with soft mooring systems adding resonance only outside the wave-frequency range.

Sometimes, time-domain simulations with non-linear static mooring reaction are performed, but wave frequency and low-frequency motion responses may alternatively be calculated separately in the frequency domain and added. In the latter case, a horizontal, linearized mooring stiffness is used. In DNV-OS-E301 the larger of the below combined horizontal offsets is thereafter used for calculation of quasi static line tension

$$X_{C1} = X_{mean} + X_{LF-max} + X_{WF-sig}$$
$$X_{C2} = X_{mean} + X_{LF-sig} + X_{WF-max}$$

$$(7.25)$$

where X_{C1} and X_{C2} are the characteristic offsets to be considered, X_{mean} is the offset caused by the mean environmental forces and, X_{LF-max} and X_{LF-sig} are, respectively, the maximum and significant offset caused by the low-frequency forces and X_{WF-max} and X_{WF-sig} the maximum and significant offset caused by the wave-frequency forces. The low- and wave-frequency motions shall be calculated in the mean offset position using the linearized mooring stiffness in the mean position. By the index max is meant the most probable maximum amplitude motion in three hours. By the index sig is meant the significant amplitude motion in three hours. If the standard deviation of motion is σ, then the significant offset is 2σ, and the most probable maximum offset is $\sqrt{0.5 \ln N} 2\sigma$ in N oscillations which means 3.72σ in 1000 waves ($T_z = 11$ s) and maybe 3σ in the slowly varying oscillations ($N = 100$, $T_z = 110$ s).

The tension caused by the greater of the two extreme offsets according to Eq. 7.25 is subsequently used to calculate the design tension in the most loaded mooring leg. For a conventional catenary system this would be in a windward mooring leg at the attachment point to the floating device.

7.5.1.2 Safety Factors

In DNV-OS-E301 two consequence classes are introduced in the ULS and ALS, defined as:

- *Class 1*, where mooring system failure is unlikely to lead to unacceptable consequences such as loss of life, collision with an adjacent platform, uncontrolled outflow of oil or gas, capsize or sinking.
- *Class 2*, where mooring system failure may well lead to unacceptable consequences of these types.

The calculated tension $T_{QS}(X_C)$ should be multiplied by a partial safety factor $\gamma = 1.7$ for Consequence Class 1 and quasi-static design from Table 7.10, and the product should be less than 0.95 times the minimum breaking strength, S_{mbs}, when statistics of the breaking strength of the component are not available:

$$\gamma T_{QS} < 0.95 S_{mbs} \tag{7.26}$$

or expressed by a utilization factor, u, which should be less than 1:

$$u = \frac{\gamma T_{QS}}{0.95 S_{mbs}} < 1 \tag{7.27}$$

Table 7.10 Partial safety factors for ULS, DNV-OS-E301 [5]

Consequence class	Type of analysis	Partial safety factor for mean tension	Partial safety factor for dynamic tension
1	Dynamic	1.10	1.50
2	Dynamic	1.40	2.10
1	Quasi-static	1.70	
2	Quasi-static	2.50	

A requirement for a slack catenary system with drag-embedment anchors is also that the mooring cables must not lift from the bottom all the way to the anchor.

Table 7.10 is quoted from DNV and contains safety factors for dynamic design, which are not used here but included for completeness.

First Design Loop

As described in Sect. 7.5.1.1 the calculated tension $T_{QS}(X_C)$ should be multiplied by a partial safety factor $\gamma = 1.7$ for Consequence Class 1 and the product should be less than 0.95 times the minimum breaking strength, S_{mbs}.

For the present example results of the design calculation are given in Table 7.11, see Sect. 7.6.8. As can be seen the calculation with the horizontal mooring stiffness $S = 12$ kN/m does not fulfil the strength requirements above, and thus we need to do a second design round with a modified mooring system.

Table 7.11 Comparison between required tension and calculated tension

Design offset (m)		Lifted chain length (m) at X_{C2}	T_{QS} (MN)	γT_{QS} (MN)	$0.95S_{mbs}$ (MN)	u
X_{C1}	X_{C2}					
7.8	12.3	424	1.38	2.35	1.9	1.23

Studless chain Q3 diameter 50.4 mm. Offset stiffness 12 kN/m

Second Design Loop

Solving Inequality 27 for the minimum breaking strength with $T_{QS} = 1.38$ MN gives a required minimum breaking strength to 2.5 MN. This corresponds e.g. to a stud chain Grade 3 diameter 58 mm [27] with $S_{mbs} = 2.6$ MN, a mass of 77 kg/m [28] and a stiffness of 296 MN [5]. A second design loop was performed with this chain and diffraction methods including linearized damping, see Tables 7.12 and 7.13. The usage factor is now 1.03 which is almost permissible. Adding 31 m to the cable gives a slightly more elastic (softer) mooring which fulfils $\gamma T_{QS} = {<}0.95\,S_{mbs}$ and the usage factor $u = 0.99 < 1$.

Table 7.12 Design offsets for quasi-static design

Mean offset	Stiffness	Wave frequency amplitude		Low frequency amplitude		Design offset		Lifted chain length at X_{C2}
(m)	(kN/m)	(m)		(m)		(m)		(m)
		Sign.	Max.	Sign.	Max.	X_{C1}	X_{C2}	
3.4	13.3	5.3	9.9	0	0	8.7	13.3	370

Diffraction results with equivalent drag damping for stud chain Grade 3 diameter 58 mm

Table 7.13 Comparison between required tension and calculated tension for stud chain Grade 3 diameter 58 mm, 509 m and 540 m long chains

Stiffness	Design offset		Lifted chain length at X_{C2}	T_{QS}	γT_{QS}	$0.95S_{mbs}$	U
(kN/m)	(m)		(m)	(MN)	(MN)	(MN)	
	X_{C1}	X_{C2}					
509 m long cable							
13.3	8.7	13.3	370	1.50	2.55	2.47	1.03
540 m long cable							
13.2	8.7	13.3	362	1.44	2.45	2.47	0.99

7.5.2 Dynamic Design

7.5.2.1 Dynamic Design Using Uncoupled Mooring Cable Dynamics

In the simplest dynamic design, the time domain motion of the attachment points of
the mooring cables is fed into some cable dynamics program to produce dynamic
forces in the cables. This is especially vital for reproducing the maximum tensions
in the cables. In Fig. 7.19 as an example, time traces of measured cable tension,
tension simulated in the cable dynamics program MODEX [29] and tension calculated from the static elastic catenary are plotted, the latter two using the measured
fairlead motion as input. One can observe that the dynamically calculated tension is
fairly close to the measured tension, while the quasistatic tension is much too
small. A similar observation was made in analyses for the WaveBob [30]. This was
often referred to as Dynamic Design around 1990. In DNV-OS-E301 [5] this is the
standard procedure for the mooring line response analysis. Programs containing this
approach are, e.g., MIMOSA [31], ORCAFLEX [32], ZENMOOR [33] and SIMO
[34]. SIMO, in combination with the cable dynamics program RIFLEX [20], has
been used by Parmeggiano et al. [35]. for the Wave Dragon.

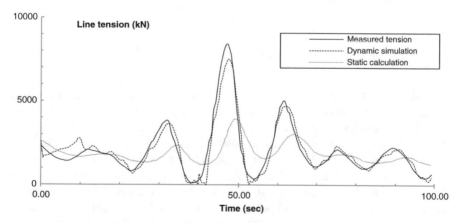

Fig. 7.19 Course of cable tension around the time for maximum tension in a model test of GVA
5000P (Troll C). [29]

7.5.2.2 Coupled Analysis

In modern computer packages for mooring design "fully" coupled mooring analysis
is often included, for example, DeepC [36], CASH [20], Orcaflex [32]. In such
analyses, the floater characteristics are first calculated in a diffraction program and
then time-domain simulations are run using convolution techniques with "full"
dynamic reaction from all mooring cables and risers. Time series of cable and riser
tensions, floater motions, air gap, etc. are output. Typically, around 10–20

realisations for each specified combination of environmental conditions (Sea state, wind speed and current velocity) are run and statistics of platform motions and cable and riser forces are subsequently evaluated. Still, the wave-induced motion is based on small-amplitude wave theory and small-amplitude body motion and viscous effects may only be included by drag formulations. This may be less inaccurate for large platforms, with moderate motions compared to their size, than for WECs. Fully coupled analysis is often used as a final check in the design, for example, for Thunder Horse [37], with a displacement of 130,000 tonnes. A fully coupled analysis of multiple wave energy converters in a park configuration is described by Gao and Moan [38], and the PELAMIS team used Orcaflex for coupled analysis of the moorings [39].

7.5.2.3   Coupled Analyses with Potential or CFD Simulations

The next natural step would be to exchange the diffraction calculation of the floating body for a non-linear potential simulation with free surface [40] or CFD RANS simulation also containing viscosity. Efforts in the latter direction for WECs are made by, for example, Palm et al. [41], and by Yu and Li [42]. Processor times are still large, but are gradually becoming more affordable.

7.5.3   Response-Based Analysis

Recently, it has become common to check the final design that was based on some specified N-year environmental combination. This is done within the framework of a "response-based analysis" using long time series of real and synthesised environmental data. For instance, such an analysis was made for the Jack & St Malo semisubmersible for Chevron [43], with 145,000 tonnes displacement, even larger than the Thunder Horse. A representative, but synthesised, 424 year period of data for every hour (3.8 million time stamps) was used as a basis. From this basis, around 380000 statistically independent "worst" events were selected. Running dynamic simulations on all these 380000 events is impractical, so these events were first screened in quasi-static analyses and around 1900 events were selected with extreme responses above specified levels. Again, the selected 1900 events were simulated by dynamic runs in the program SIMO using a somewhat simplified input for current drag and viscous effects. Of the 1900 events, around 220 met higher extreme response levels. Finally, these 220 events were simulated in SIMO with an updated current drag model calibrated against model tests for each sea state. In a statistical analysis, the N-year response was calculated and compared to the responses of the N-year environmental design combinations. In this case, the responses to the N-year design conditions were found to be worse or equal to the simulated N-year responses for both 100 and 1000 year return periods [44].

It may be anticipated that in the future response-based analysis could be used for a last check of the design of ocean energy converters.

7.6 Response Motion of the Moored Structure

7.6.1 Equation of Motion

The forces on a floating body can be constant as the mean force in Sect. 7.4.3, transient i.e. of short duration or harmonic. Irregular or random forces from e.g. sea waves can to a first, linear approximation be treated as a superposition of harmonic forces, an approach that will be used here. The responses are fundamentally different for the three types of forces. The present buoy—mooring system will be treated as a single-degree-of-freedom (SDoF) system as illustrated in Fig. 7.20.

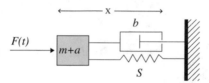

Fig. 7.20 A mechanical system with one degree of freedom, mass, m, added mass, a, damping coefficient, b, and spring stiffness, S

The equation of motion for this system can be written

$$(m+a)\ddot{x} + b\dot{x} + Sx = F(t) \tag{7.28}$$

For bodies in water the mass inertia is increased by an "added mass", a, or hydrodynamic mass. In our case this is represented by the C_m coefficient. This is a result of the fact that to accelerate the body it is also necessary to accelerate the water surrounding the body. For submerged bodies close to the water surface the added mass can be negative, but for deeply submerged bodies it is always positive. For bodies vibrating in or close to the water surface the damping, b, is caused by the radiation of waves from the motion of the buoy and also by linearized viscous damping through the drag force. The coefficients a and b are functions of the motion frequency, or wave frequency in waves, see e.g. Figs. 7.9 and 7.10 for the sample buoy. S is the mooring stiffness and $F(t)$ is the driving force.

General mechanics of vibration can be found in some fundamental textbooks e.g. books by Craig [45], Roberts and Spanos [46] or Thompson [47].

7.6.2 Free Vibration of a Floating Buoy in Surge

Before the discussion of response to different types of forcing we will repeat a little about the free vibrations of the one-degree-of-freedom system. The equation of motion for a buoy in surge can be written

$$(m+a)\ddot{x} + b\dot{x} + Sx = 0 \tag{7.29}$$

which follows directly from Eq. 7.28 setting $F(t) = 0$.

Assuming a solution of the form

$$x = Ce^{\kappa t} \tag{7.30}$$

we get the characteristic equation

$$\kappa^2 + 2\xi\omega_N\kappa + \omega_N^2 = 0 \tag{7.31}$$

where

- $\omega_N = \sqrt{S/(m+a)}$ is the "natural" angular frequency, that is, the undamped angular frequency
- $\xi = b/\left(2\sqrt{S(m+a)}\right)$ is the damping factor.

The roots of $\kappa^2 + 2\xi\omega_N\kappa + \omega_N^2 = 0$ $\tag{7.32}$

are

$$\kappa_{1,2} = -\xi\omega_N \pm \omega_N\sqrt{\xi^2 - 1}. \tag{7.33}$$

These roots are complex, zero or real depending on the value of ξ. The damping factor can thus be used to distinguish between three cases: underdamped ($0 < \xi < 1$), critically damped ($\xi = 1$) and overdamped ($\xi > 1$). See Fig. 7.21 for the motion of a body released from the position $x(0) = 1$ m at $t = 0$ s. The underdamped case displays an attenuating oscillation, while the other cases display motions monotonously approaching the equilibrium position. A moored floating buoy in surge would normally display underdamped characteristics with a damping factor of the order of 10^{-3}. Note that an unmoored buoy, $S = 0$ exhibits no surge resonance. The damping factor is often called the damping ratio, as it is equal to the ratio between the current damping coefficient, b, and the critical damping coefficient, $2\sqrt{c(m+a)}$.

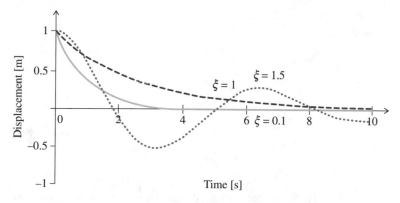

Fig. 7.21 Response of a damped SDOF system with various damping ratios

The natural frequency and damping factor for the moored buoy at the mean offset are listed in Table 7.14. As the peak period is $T_p = 12.9$ s and the zero-crossing period is $T_{02} = 9.2$ s $< T_z < T_{02} = 9.9$ s in the design spectrum, there is a risk for horizontal resonant motion.

Table 7.14 Natural frequencies and damping factors for the moored buoy at the mean offsets

Mean offset (m)	Stiffness (kN/m)	Natural period (s)	Damping factor
2.6	12.0	24.4	0.09×10^{-3}

7.6.3   Response to Harmonic Forces

A harmonic force

$$F(t) = F_o\cos(\omega t) \tag{7.34}$$

as from regular waves for instance gives a response of the same harmonic type:

$$x(t) = \hat{x}\ \cos(\omega t - \varepsilon). \tag{7.35}$$

The motion $x(t)$ is the stationary response to the harmonic force and is the particular solution to Eq. 7.28 with the right hand side $F(t)$ given by Eq. 7.34

F_o is the force amplitude

$\omega = 2\pi/T$ the angular frequency T the time period

$\hat{x}$ the amplitude of the displacement and ε the phase lag between the force and displacement (Fig. 7.22).

We can solve

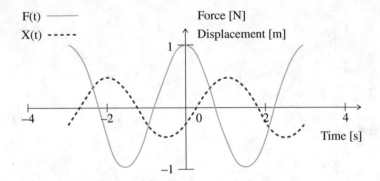

Fig. 7.22 The exciting harmonic force $F(t)$ and the stationary Response, $x(t)$, for a linear system

$$(m+a)\ddot{x} + b\dot{x} + Sx = F(t) \tag{7.28}$$

for the given harmonic force,

$$F(t) = F_o\cos(\omega t) \tag{7.34}$$

simply by substituting the particular solution Eq. 7.35 into it. The last equation gives the surge velocity and acceleration of the buoy:

$$x = \hat{x}\cos(\omega t - \varepsilon)$$
$$\dot{x} = -\omega\hat{x}\sin(\omega t - \varepsilon)$$
$$\ddot{x} = -\omega^2\hat{x}\cos(\omega t - \varepsilon)$$

The substitution gives

$$(S - (m+a)\omega^2)\hat{x}\cos(\omega t - \varepsilon) - b\omega\hat{x}\sin(\omega t - \varepsilon) = F_o\cos(\omega t) \tag{7.36}$$

Using the trigonometric expressions for sine and cosine of angle differences then after some manipulation yields the amplitude $\hat{x}$, which by definition is positive.

$$\hat{x} = \frac{F_o}{\sqrt{\left\{(S - (m+a)\omega^2)^2 + b^2\omega^2\right\}}} \tag{7.37}$$

We can solve for the phase angle, ε, also, but this is not of interest in the present context. In Fig. 7.23 a graph is drawn of the horizontal response amplitude ratio, i.e. the surge motion amplitude divided by the wave force amplitude, as a function of frequency. The frequencies corresponding to the peak and mean periods are marked to point out the sensitivity to the forcing frequency.

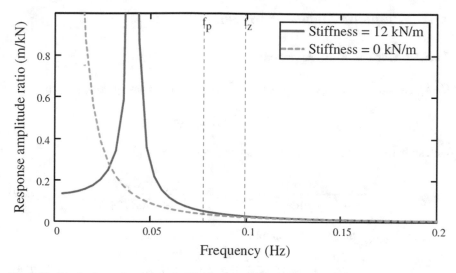

Fig. 7.23 The horizontal response amplitude ratio, surge motion amplitude divided by the wave force amplitude, as a function of frequency. The frequencies corresponding to the peak and mean periods of the wave spectrum are marked to point out possible effects of the forcing frequency

In Table 7.15 the amplitude of the excursion around the mean offset is listed for a regular waves with the periods T_p and $T_z = T_{01}$ with the force amplitude $F_o = F_{Msamp} = 0.38$ MN, i.e. the value of significant force amplitude. In the case of a fixed structure the maximum wave would produce the largest force on the structure. However, for the motion of a moored structure Eq. 7.36 gives the asymptotic motion amplitude after several regular force cycles, while the maximum wave just is a transient incident. It may therefore be more appropriate to use the significant wave height, combined with the peak or mean period. Because the system is very sensitive to resonance, we need include drag damping in a time-domain model or at least linearized drag damping to get near realistic results. Note that the motion amplitude of an unmoored buoy exhibits a smaller motion amplitude due to the absence of resonance.

Table 7.15 Motion amplitude due to a regular Morison wave force, $F_0 = F_{Msamp} = 0.38$ MN

Stiffness (kN/m)	Amplitude at f_p (m)	Amplitude at f_z (m)	Mean offset (m)	Combined excursion (m)	
				at f_p	at f_z
12	12.3	6.5	2.6	14.9	9.1
Unmoored	8.8	5.4		8.8	5.4

7.6.4   Response Motion in Irregular Waves

7.6.4.1   Morison Mass Approach

Using the wave force spectrum based on the Morison mass force approach

$$S_F(f) = (f_w(f))^2 S_{PM}(f), \tag{7.38}$$

we can calculate the surge motion response spectrum as [22]

$$S_x(f) = \frac{S_F(f)}{(S - (m+a)\omega^2)^2 + b^2\omega^2} = \frac{(f_w(f))^2 S_{PM}(f)}{(S - (m+a)\omega^2)^2 + b^2\omega^2} \tag{7.39}$$

Then the significant motion amplitude can be estimated as

$$x_{1s} = 2\sqrt{m_{0dF}} = 2\sqrt{\sum_i S_x(f_i)\Delta f_i} \tag{7.40}$$

The result of this calculation is shown in Fig. 7.24 and in Table 7.16 below on the lines marked "none" under linearized drag damping. Without consideration of the drag damping the motion becomes unrealistically large as the large horizontal drag damping is not taken into account. It is much larger than the surge radiation damping.

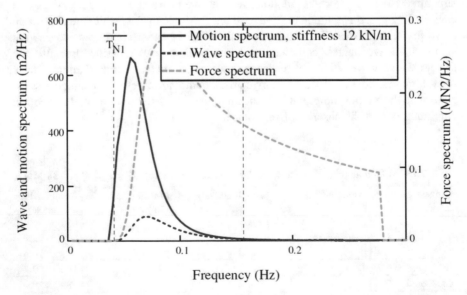

Fig. 7.24 Motion spectra, wave spectrum and force spectrum as functions of frequency. Morison mass approach. No viscous damping. The natural motion period is also marked as $1/T_{N1}$

Table 7.16 Significant linear response in an irregular wave, PM-spectrum, $H_s = 8.3$ m

Mean offset (m)		Stiffness (kN/m)	Linearized drag damping	Significant amplitude (m)
2.6	Morison	12	None	7.3
2.6	Diffraction	12	None	9.5
2.6	Diffraction	12	Included	5.2

7.6.4.2   Diffraction Force Approach

Using the wave force spectrum based on diffraction forces we can similarly form a diffraction-based surge spectrum:

$$S_{dF}(f) = (f_{dw}(f))^2 S_{PM}(f) \tag{7.41}$$

we can calculate the surge motion response spectrum as [22]

$$S_{dx}(f) = \frac{S_{dF}(f)}{(S - (m+a)\omega^2)^2 + b^2\omega^2} = \frac{(f_{dw}(f))^2 S_{PM}(f)}{(S - (m+a)\omega^2)^2 + b^2\omega^2} \tag{7.42}$$

Then the significant motion amplitude can be estimated as

$$x_{d1s} = 2\sqrt{m_{0dF}} = 2\sqrt{\sum_i S_x(f_i)\Delta f_i} \tag{7.43}$$

The result of this calculation is shown in Fig. 7.25 and in Table 7.16 on the lines marked diffraction and "none" under linearized drag damping. Without consideration of the drag damping the motion becomes also here unrealistically large.

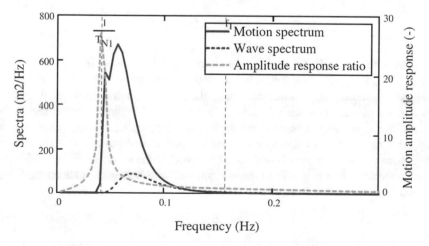

Fig. 7.25 Motion spectrum and wave spectrum as functions of frequency for the stiffness 12 kN/m. The natural frequency $1/T_{N1}$ is marked. Diffraction approach. No viscous damping

7.6.5   Equivalent Linearized Drag Damping

Neglecting the coupling between surge and pitch we can symbolically write the drag damping surge force as

$$F_{D1} = K|u - \dot{x}_1|(u - \dot{x}_1), \tag{7.44}$$

where K can be set to $(1/2)\rho C_D D h_b$ and u is the undisturbed horizontal velocity of the water in the surge direction and $\dot{x}_1$ the surge velocity of the buoy.

When the non-linear surge damping is important usually $u \ll \dot{x}_1$ and then we can set

$$F_{D1} = K|\dot{x}_1|(\dot{x}_1), \tag{7.45}$$

which is simpler but still non-linear.

To assess an equivalent linear coefficient we can compare the dissipated energy over a time, say 3 h, with an equivalent linear expression and the surge velocity

$$x_1(t) = \sum_i \left(\sqrt{2S_x(f_i)\Delta f_i} \cos(\omega_i t + \varepsilon_i) \right) \tag{7.46}$$

Then the dissipated energy can be calculated in two ways

$$\int_0^T K|\dot{x}_1|(\dot{x}_1)^2 dt = \int_0^T B_{e11}(\dot{x}_1)^2 dt, \tag{7.47}$$

$$\therefore B_{e11} = K \frac{\int_0^T |\dot{x}_1|(\dot{x}_1)^2 dt}{\int_0^T (\dot{x}_1)^2 dt} \tag{7.48}$$

That is, the equivalent damping coefficient, B_{e11}, depends on the modulus of the surge motion, $|\dot{x}_1|$. The result of this calculation is shown in Table 7.16 on the line marked "included" under linearized drag damping. It should be warned that the specific set of wave components and phase angles used in the numerical realisation affects the equivalent damping and significant amplitudes. In our case we got around 8 m significant amplitude for one realisation and around 5 m for another one. However, we may now be able to accommodate the motion. In Fig. 7.26, there is a comparison between surge response spectra with and without linearized drag damping.

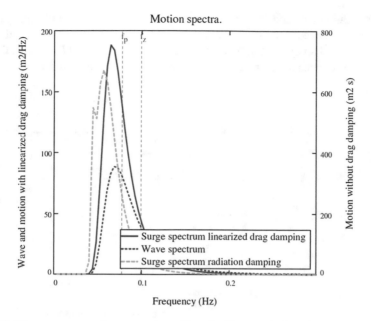

Fig. 7.26 Wave spectrum and surge spectra with and without equivalent linearized damping for the stiffness 12 kN/m. Note the different vertical scales

7.6.6   Second-Order Slowly Varying Motion

In cases where the second-order slowly varying wave force hits the resonance of the moored system, second order slowly varying motion may become large and induce motions of the same order of magnitude as the first-order wave induced motions.

The low-frequency excitation force can be expressed in the frequency-domain by a spectrum [48].

$$S_{LF}(\mu) = 8 \int_0^\infty S(\omega)S(\omega+\mu)C_d\left(\omega+\frac{\mu}{2}\right)d\omega \qquad (7.49)$$

Here $S(\omega)$ is the wave spectrum and $C_d(\omega)$ is the wave drift force coefficient. The equation is invoking the Newman [49] approximation and cannot be used if the resonance period is within the wave spectrum periods. Then the full non-linear expression should be used, see e.g. [15]. In the present case this is not the case and, anyway, in such cases the motion is dominated by the first-order wave-excited motion.

A sample calculation for this case gives negligible second order slowly varying motion—surge amplitude in the order of mm—compared to the first-order motion. The first and second order motions can be comparable in lower sea states. The reason for negligible second order slowly varying motion is that the resonance period is off the peak of the drift force spectrum and that the drift force coefficient is small. On the other hand, we should maybe have used the full non-linear expression. However, experience gives that the second-order motions for small objects in high sea states display little second-order motions. See Fig. 7.27, where a range of horizontal resonance angular frequencies from 0.3 to 0.6 rad/s for realistic offsets is marked.

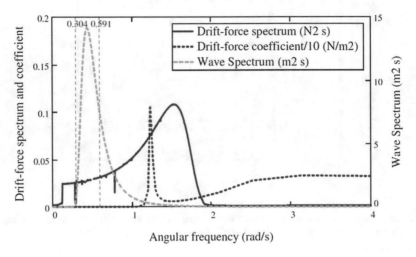

Fig. 7.27 Drift-force spectrum, drift-force coefficient and wave spectrum as functions of angular frequency

7.6.7 Wave Drift Damping

In forward speed and in coastal currents the slowly varying motion may be damped by the fact that the encountered wave period and subsequently the wave drift coefficient varies during the slow surge causing a kind of hysteretic damping, called wave-drift damping. As we have negligibly small slowly varying motion in the present case, it is not useful to take this into account.

7.6.8 Combined Maximum Excursions

Using the design format according to Sect. 7.5.1.1 we end up with the following table over the design motions X_{C1} and X_{C2} (Table 7.17).

Table 7.17 Design offsets for quasi-static design

Mean offset	Stiffness							Low-frequency amplitude
Wave-frequency amplitude								Low-frequency amplitude
Design offset								Lifted chain length at X_{C2}
(m)	(kN/m)	(m)		(m)		(m)		(m)
		Sign.	Max.	Sign.	Max.	X_{C1}	X_{C2}	
2.6	12	5.2	9.7	0	0	7.8	12.3	424

Diffraction results with equivalent drag damping

The calculation shows that if we use the tangent stiffness modulus (12 kN/m) of the mooring system we fulfil the lifting criterion that the up-wave chain should rest on the bottom close to the anchor. However, in Sect. 7.5.1.2.1 it is shown that we do not fulfil the tension criterion, why a second design loop was performed.

7.7 Conclusions

The following conclusions can be drawn from the design exercise

- Simplified drag and wind coefficients can be used, because the mean offset is not a dominant part of the total horizontal displacement.
- The Morison wave formulation can be used for objects smaller than 1/4th of the wavelength, however with some overdesign. It is important to test various wave frequencies and realistic wave amplitudes. Used in the frequency-domain equivalent linearized drag damping must be added to compensate for the dropped drag term.
- Using the diffraction method for small objects, equivalent linearized drag damping must be added.
- In the equation of motion, there is a difficulty with progressively stiffening moorings. In the CALM system choosing a stiffness around the mean offset will not give a realistic motion as the stiffness may vary one order of magnitude during the oscillation. It is advised to use time-domain simulations taking at least $S(x)$ into consideration, and then the drag damping could as well be introduced as $b(\dot{x}) = CA\,1/2\,\rho|\dot{x}|$.
- In a final design, time-domain design tools including mooring dynamics should be used complemented by large-scale model tests.
- Other types of moorings as e.g. synthetic fibre ropes in taut configuration or with buoys and lump weights may better fulfil demands on footprint and non-resonant motions. The weight of the chains may cause a large vertical force on the floater, which may constitute a problem. For taught systems the anchoring must take vertical lifting forces, which must be studied.

Acknowledgments The underlying study was carried out at Dept. of Shipping and Marine Technology, Chalmers, and was co-funded from Region Västra Götaland, Sweden, through the Ocean Energy Centre hosted by Chalmers University of Technology, and the Danish Council for Strategic Research under the Programme Commission on Sustainable Energy and Environment (Contract 09-067257, Structural Design of Wave Energy Devices).

References

1. Bergdahl, L., McCullen, P.: Development of a Safety Standard forWave Power Conversion Systems. Wave Energy Network, CONTRACT N°: ERK5 - CT - 1999-2001 (2002)
2. Bergdahl, L., Mårtensson, N.: Certification of wave-energy plants—discussion of existing guidelines, especially for mooring design. In: Proceedings of the 2nd European Wave Power Conference, pp. 114–118. Lisbon, Portugal (1995)
3. Johanning, L., Smith, G.H., Wolfram, J.: Towards design standards for WEC moorings. In: Proceedings of the 6th Wave and Tidal Energy Conference, Glasgow, Scotland (2005)
4. Paredes, G., Bergdahl, L., Palm, J., Eskilsson, C., Pinto, F.: Station keeping design for floating wave energy devices compared to floating offshore oil and gas platforms. In: Proceedings of the 10th European Wave and Tidal Energy Conference 2013
5. Position Mooring, DNV Offshore Standard DNV-OS-E301, 2013
6. Guidelines on design and operation of wave energy converters, Det Norske Veritas, 2005 (Carbon Trust Guidelines)
7. SWAN, Simulating Waves Nearshore. http://swanmodel.sourceforge.net/. Accessed 10 Aug 2015
8. http://www.mikepoweredbydhi.com/-/media/shared%20content/mike%20by%20dhi/flyers%20and%20pdf/product-documentation/short%20descriptions/mike21_sw_fm_short_description.pdf. Accessed 19 Aug 2015
9. Recommended Practice DNV-RP-C205, October 2013
10. Long-term Wave Prediction, Structural Design of Wave Energy Devices, Deliverable D1.1, DHI, Erwan Tacher, Jacob V Tornfeldt Sørensen, Maziar Golestani, 10 April 2013
11. Pecher, A., Kofoed, J.P.: Experimental study on the structural and mooring loads of the WEPTOS Wave Energy Converter. Aalborg: Department of Civil Engineering, Aalborg University. (DCE Contract Reports; No. 142) (2014)
12. Sterndorf, M.: WavePlane, Conceptual mooring system design, Sterndorf Engineering, 15 November 2009
13. Margheritini, L.: Review on available information on wind, water level, current, geology and bathymetry in the DanWEC area, (DanWEC Vaekstforum 2011), Dep. Civil Engineering, Aalborg University, Aalborg, DCE Technical Report, No 135 (2012)
14. Pecher, A., Foglia, A., Kofoed, J.P.: Comparison and sensitivity investigations of a CALM and SALM type mooring system for WECs
15. Faltinsen, O.M.: Sea Loads on Ships and Offshore Structures. Cambridge University Press (1990)
16. Sachs, P.: Wind Forces in Engineering, 2nd edn. Elsevier (1978) ISBN: 978-0-08-021299-9
17. Haddara, M.R., Guedes, C.: Soares: wind loads on marine structures. Marine Struct. **12**, 199–209 (1999)
18. Longuet-Higgins, H.C.: The mean forces exerted by waves on floating or submerged bodies with application to sand bars and wave-power machines. Proc. R. Soc. London **A352**, 462–480 (1977)
19. Maruo, H.: The drift of a body floating on waves. J. Ship Res. **4** (1960)
20. Sesam, DeepC, DNV Softwares. http://www.dnv.com/services/software/products/sesam/sesamdeepc/. Accessed 03 Nov 2013

21. Sarpkaya, T., Isaacson, M.: Mechanics of Wave Forces on Offshore Structures, Van Nostrand Reinhold Company (1981) ISBN 10: 0442254024/ISBN 13: 9780442254025
22. Chakrabarti, S.K.: Handbook of offshore engineering, vol. 1. Elsevier (2005)
23. Yeung, R.: Added mass and damping of a vertical cylinder in finite-depth waters. Appl. Ocean Res. 3(3), 119–133 (1981) ISSN 0141-1187
24. Johansson, M.: Transient motion of large floating structures, Report Series A:14, Department of Hydraulics, Chalmers University of Technology (1986)
25. Ramsey, A.S.: Statics. The University Press, Cambridge (1960)
26. Selmer, J.: Forankrings- og fortøyningssystemer. Beskrivelse av analysmetodikk og belastningforhold. Course: Kjetting, ståltau og fibertau, Norske sivilingenjørers forening, 1979
27. http://www.sotra.net/products/tables/stength-for-studlink-anchor-chain-cables. Accessed 05 May 2014
28. http://www.sotra.net/products/tables/weight-for-studlink-anchor-chain. Accessed 22 May 2014
29. Bergdahl, L.M., Rask, I.: Dynamic vs. Quasi-Static Design of Catenary Mooring System, 1987 Offshore Technology Conference, OTC 5530
30. Muliawan, M.J., Gao, Z., Moan, T.: Analysis of a two-body floating wave energy converter with particular focus on the effect of power take off and mooring systems on energy capture. In: OMAE 2011, OMAE2011-49135
31. Sesam, MIMOSA, DNV Softwares. http://www.dnv.com/services/software/products/sesam/sesamdeepc/mimosa.asp. Accessed 03 Nov 2013
32. Orcaflex Documentation. http://www.orcina.com/SoftwareProducts/OrcaFlex/Documentation/OrcaFlex.pdf. Accessed 03 Nov 2013
33. ZENMOOR Mooring Analysis Software for Floating Vessels. http://www.zentech-usa.com/zentech/pdf/zenmoor.pdf. Accessed 03 Dec 2012
34. Sesam SIMO, DNV Softwares. http://www.dnv.com/services/software/products/sesam/sesamdeepc/simo.asp. Accessed 03 Nov 2013
35. Parmeggiano, S. et al.: Comparison of mooring loads in survivability mode of the wave dragon wave energy converter obtained by a numerical model and experimental data. In: OMAE 2012, OMAE2012-83415
36. CASH, In-house program, GVA. http://www.gvac.se/engineering-tools/. Accessed 03 Nov 2013
37. Thunder Horse Production and Drilling Unit. http://www.gvac.se/thunder-horse/. Accessed 03 Nov 2013
38. Gao, Z., Moan, T.: Mooring system analysis of multiple wave energy converters in a farm configuration. In: Proceedings of the 8th European Wave and Tidal Energy Conference, Uppsala, Sweden (2009)
39. Pizer, D.J. et al.: Pelamis WEC—Recent advances in the numerical and experimental modelling programme. In: Proceedings of the 6th European Wave and Tidal Energy Conference, Glasgow, UK (2005)
40. Ma, Q.W., Yan, S.: QALE-FEM for numerical modelling of nonlinear interaction between 3D moored floating bodies and steep waves. Int. J. Numer. Meth. Engng. 78, 713–756 (2009)
41. Palm, J. et al.: Coupled mooring analysis for floating wave energy converters using CFD: Formulation and validation. Int. J. Mar Energy (May 2016). doi: 10.1016/j.ijome.2016.05.003
42. Yu, Y., Li, Y.: Preliminary result of a RANS simulation for a floating point absorber wave energy system under extreme wave conditions. In: 30th International Conference on Ocean, Offshore, and Arctic Engineering, Rotterdam, The Netherlands, 19–24 June 2011
43. Jack/St Malo Deepwater Oil Project. http://www.offshore-technology.com/projects/jackstmalodeepwaterp/. Accessed 03 Nov 2013
44. Jack & St Malo Project, Response-based analysis, GVA, KBR, Göteborg, 2010, Internal report
45. Craig Roy, Jr. R.: Structural Dynamics. Wiley, New York (1981)

46. Roberts, J.B., Spanos, P.D.: Random vibration and statistical linearization. Wiley, Chichester (1990)
47. William, T.: Thompson: Theory of Vibration with Applications. Prentice-Hall Inc., Englewood Cliffs (1972)
48. Pinkster, J.A.: Low-frequency phenomena associated with vessels moored at sea. Soc. Petrol. Eng. J. pp. 487–494 (1975)
49. Newman, J.N.: Second Order Slowly Varying Forces on Vessels in Irregular Waves. In: Bishop, R.E.D, Price, W.G. (eds.) Proceedings of International Symposium Dynamics of Marine Vehicles and Structures in Waves. pp. 182–186. London Mechanical Engineering Publications LTD (1974)

Chapter 8
Power Take-Off Systems for WECs

Amélie Têtu

8.1 Introduction, Importance and Challenges

The power take-off (PTO) of a wave energy converter is defined as the mechanism with which the absorbed energy by the primary converter is transformed into useable electricity. The primary converter can for example be an enclosed chamber for an oscillating water column or a point absorber buoy. The PTO system is of great importance as it affects not only directly how efficiently the absorbed wave power is converted into electricity, but also contributes to the mass, the size and the structural dynamics of the wave energy converter.

By having this direct influence on the wave energy converter, the PTO system has a direct impact on the levelised cost of energy (LCoE) [1]. The PTO system has a direct effect on the efficiency of power conversion; hence, it has a direct impact on the annual energy production. The PTO system affects directly the capital cost of a device by accounting for typically between 20–30 % of the total capital cost [2]. The reliability of the PTO system affects the availability (the energy production) and the operation and maintenance cost.[1] The influence of the PTO on the LCoE is schematized in Fig. 8.1. A study made by the Partnership for Wave Energy in Denmark investigated the influence of the PTO system for four different wave energy converters [3]. The impact of the PTO efficiency and the reduction in cost of the PTO system on the LCoE were the PTO variables studied; Fig. 8.2 shows the results.

---

[1]For more information on those economic variables, the reader is referred to Chap. 5 of this book.

A. Têtu (✉)
Department of Civil Engineering, Aalborg University, Thomas Manns Vej 23, 9220 Aalborg
Ø, Denmark
e-mail: at@civil.aau.dk

© The Author(s) 2017
A. Pecher and J.P. Kofoed (eds.), *Handbook of Ocean Wave Energy*,
Ocean Engineering & Oceanography 7, DOI 10.1007/978-3-319-39889-1_8

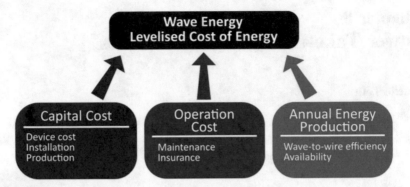

Fig. 8.1 Economic variables defining the levelised cost of energy for wave energy converters. The PTO system has a direct impact on the capital cost, the operation cost and the annual energy production of the device [1]

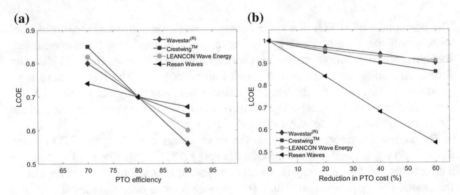

Fig. 8.2 Influence of **a** the PTO efficiency and **b** the relative reduction (in %) in the cost of the PTO system on the relative LCoE for different wave energy converters

For both an increase in efficiency and a reduction in cost of the PTO, a decrease of the LCoE is observed. Even though an increase in PTO efficiency has a bigger effect on the LCoE, both parameters have a significant impact on the LCoE showing the importance of the PTO system in a wave energy converter.

But the task of designing a cost-efficient PTO system is definitely not an easy one. The main challenge comes from the intrinsic properties of the energy resource. Ocean energy presents high variability. As shown in Fig. 8.3, the surface elevation varies irregularly in time and can induces high amplitude displacements, accelerations and forces on a body in a very short period of time. At other instants, the waves present low amplitude displacements, accelerations and forces. Those two extreme regimes present different dynamic load patterns and in both cases, the PTO system should be as efficient as possible.

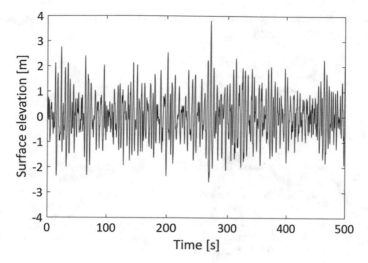

Fig. 8.3 Surface elevation as a function of time

WECs are placed in very harsh environment, leading to a high wear-rate and are difficult to access due to their location and/or unfavourably weather conditions. As for the rest of the device, the PTO system should be robust, reliable and should require as little maintenance as possible.

As opposed to the wind energy sector, there is no industrial standard device for wave energy conversion and this diversity is transferred to the PTO system. Many different types of PTO systems have been investigated, and the type of PTO system used in a wave energy converter is often correlated with its type. For example, oscillating water column type of device utilised an air turbine coupled to the electrical generator, while point absorber type of converter can use different PTO systems depending on their configuration and may require cascaded conversion mechanisms. This variety means that PTO systems are still at the development stage with little experience gained for large scale devices. To add to the difficulties, PTO systems are difficult to test at small scale as friction becomes an issue. They can first be tested at a larger scale where costs are significantly increased.

The PTO system is a crucial component of a wave energy converter. As previously mentioned, it is also difficult to design due to the variability of the energy source, the environment in which it is placed and scaling issues. This chapter aims first at giving an overview of the different types of PTO systems. The concept of control and its importance for PTO systems will then be introduced.

8.2   Types of Power Take-Off System

8.2.1   Overview

As mentioned earlier, many different types of PTO systems exist, and the type of PTO chosen for a particular wave energy converter is often strongly correlated with the type of converter. The different main paths for wave energy to electricity conversion are schematised in Fig. 8.4.

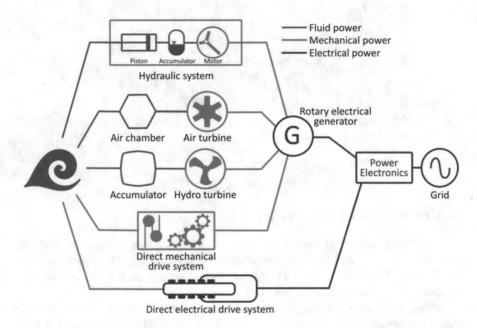

Fig. 8.4 Different paths for wave energy to electricity conversion

A systematic comparison of the different types of PTO is a difficult task to accomplish as limited data is available and one particular device can be bound to only two types of systems. The types of PTO systems can be categorized into five main categories and are described in the following sections.

8.2.2 Air Turbines

Air turbines as a mean for converting wave power into mechanical power are mostly used in oscillating water column (see Chap. 2). The idea is to drive a turbine with the oscillating air pressure in an enclosed chamber as a consequence of the oscillating water level, induced by the ocean waves (see Fig. 8.5). The main challenge comes from the bidirectional nature of the flow. Non-returning valves to rectify the air flow combined with a conventional turbine is one solution. However, this configuration is complicated, has high maintenance cost and for prototype size the valves become too large to be a viable option. Another solution is to use a self-rectifying air turbine that converts an alternating air flow into a unidirectional rotation.

Fig. 8.5 Schematic of a
wave energy converter where
an air turbine is employed

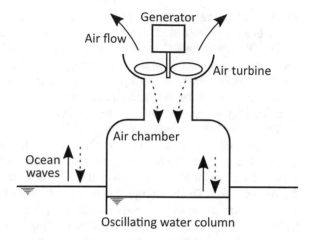

Several types of self-rectifying turbines have been proposed in the last 40 years, and new ideas are still being pursued to find an efficient reliable PTO system for the OWC systems, the main ones being:

- Wells type turbines
- Impulse turbines
- Denniss-Auld turbines

Wells type turbine was the first self-rectifying turbine to be developed and is named after its inventor A.A. Wells. It consists of a symmetrical rotor composed of many aerofoil blades positioned around a hub with the normal of their chords planes aligned with the axis of rotation (see Fig. 8.6a). When the rotor is in movement, the rotational speed induces an apparent flow angle α, which in turn creates a lift force perpendicular to the apparent flow direction and a drag force parallel to the apparent flow direction (see Fig. 8.6b). Those forces can be decomposed into axial (F_x) and tangential force (F_θ). For some given value α, the

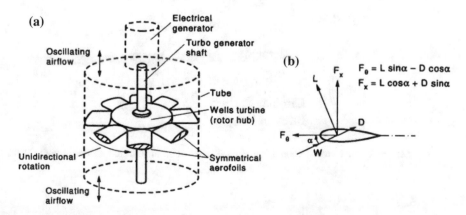

Fig. 8.6 Illustration of a self-rectifying Wells turbine (taken from [37])

direction of the tangential force is independent of the sign of α, and the rotor will rotate in a single direction regardless of the direction of the air flow.

The Wells turbine is the simplest of all the self-rectifying turbines and probably the most economical option for wave energy conversion. The Azores Pico Plant [4] and the LIMPET in Islay, U.K., [5] are both equipped with this type of turbine. One major drawback of the Wells turbines is that they are not self-starting: the rotor has to be initially accelerated by an external source of energy.

To overcome the drawbacks of the Wells turbine, the so called impulse turbine was developed. The idea is to redirect the air flow by using guide vanes in order to directly transfer the kinetic energy of the air flow into the tangential force component on the rotor blades, as depicted in Fig. 8.7. The guide vanes can either be fixed or pitched. The pitching mechanism can either be self-controlled by the air flow or controlled by another active mechanism, for example hydraulic actuator [6]. This extra feature increases the amount of moving parts of the turbine and therefor decreases the reliability and increases the operation and maintenance cost of the turbine. On the other hand, the pitching mechanism increases considerably the efficiency of the turbine, cf. Fig. 8.8.

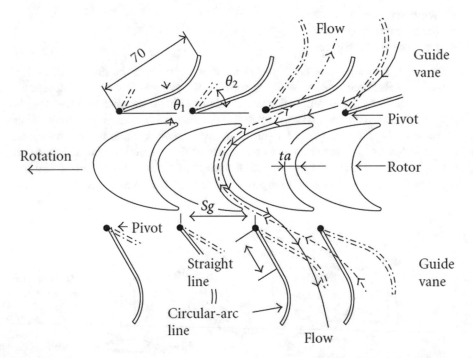

Fig. 8.7 Schematic of the cross-section at the aerofoils level of an impulse turbine (taken from [7])

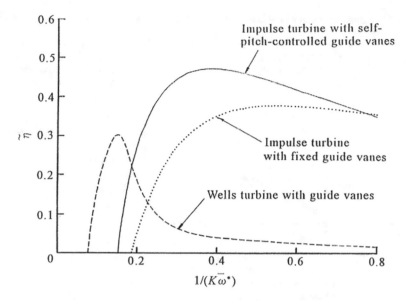

Fig. 8.8 Comparison of efficiency for different self-rectifying air turbines under irregular flow conditions (taken from [38])

An impulse turbine with self-pitch-controlled guide vanes was tested at the NIOT plant in India and showed a threefold increase in total efficiency with respect to the previously installed Wells turbine [7].

The Denniss-Auld self-rectifying turbine is based on the Wells turbines configuration where the aerofoils blades can rotate around their neutral position in order to achieve optimal angle of incident flow (see Fig. 8.9). The rotation of the blades is controlled by measuring the pressure in the chamber. This type of turbine was installed in the MK1 OWC full-scale prototype deployed in New South Wales, Australia in 2005 [8].

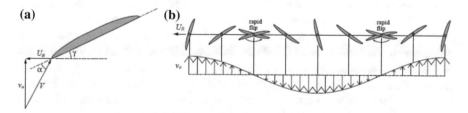

Fig. 8.9 a Schematic of an aerofoil and **b** illustration of the aerofoil pitching sequence in oscillating flow for a Denniss-Auld turbine (taken from [39])

Table 8.1 summarises the technological advantages and inconvenients for the self-rectifying air turbines mentioned above.

Table 8.1 Advantages and inconvenients for different self-rectifying turbine employed in wave energy conversion

Turbine type	Advantages	Inconvenients
Wells turbine	• Technologically simple	• Narrow flow range at which the turbine operates at useful efficiencies
		• Poor starting characteristics
		• High operational speed and consequent noise
		• High axial thrust
Impulse turbines	• Good starting characteristics	• Large number of movable parts for the self-pitching configuration
	• Low operational speed	
	• Wide range of flow coefficients at which the turbine operates at useful efficiencies	
Denniss-Auld turbines	• Low operational speed	• Large number of movable parts
	• Wide range of flow coefficients at which the turbine operates at useful efficiencies	

8.2.3   Hydraulic Converters

When the energy capture mechanism is based on the movement of a body in response to the interaction with the waves, as is the case for some point absorbers and attenuators, conventional rotary electrical machines are not directly compatible. Hydraulic converter is often the solution chosen to interface the wave energy converter with the electrical generator since they are well suited to absorb energy when dealing with large forces at low frequencies. In this particular case, the energy path is usually reversed with respect to traditional hydraulic system. The movement of the body is feeding energy into a hydraulic motor, which in turn translates the energy to an electrical generator.

A schematic of a hydraulic PTO system for wave energy conversion is depicted in Fig. 8.10. A point absorber connected to an hydraulic cylinder moves up and down with respect to an actuator, forcing fluid through controlled hydraulic manifolds to a hydraulic motor, which in turn drives the electric generator. Accumulators are also added to the system so as to smoothen the supply of high pressure fluid in the system by either providing or accumulating hydraulic energy when necessary. For wave energy conversion, radial piston motor is often favoured as it is well suited for high loads, low velocity applications.

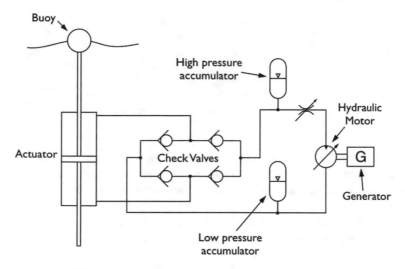

Fig. 8.10 Example of a hydraulic PTO system for wave energy conversion (taken from [40])

Many issues arise when choosing a hydraulic PTO system for wave energy conversion. Fluid containment of the hydraulic system has to be addressed with regards to performance and environmental impacts. The use of biodegradable transformer oil has been reported to address the environmental issue [9]. Efficiency of the whole system is also of importance. Due to the variability of the energy resource, hydraulic systems often include several hydraulic gas accumulators that can store the absorbed peak loads and smoothen the wave energy conversion from the motor. Digital displacement motors [10], based on radial piston motor, were developed in order to increase the part-load efficiency of hydraulic motor and facilitate their controllability [11]. Hydraulic systems are composed of many moving parts, and the seals of the piston will wear over time which can increase drastically the maintenance cost. This has to be kept in mind while designing a hydraulic PTO system. Another issue to address is the protection of the PTO system in the event of extreme conditions, where the hydraulic actuator exceeds its design travel and damage the system. One solution is to include mechanical limit to the stroke [12] or to use radial hydraulic piston [13].

8.2.4 Hydro Turbines

Hydro turbines are employed in overtopping devices or hydraulic pump systems using seawater as fluid. In overtopping type of devices, the water reaching over a ramp accumulates in a basin, and its potential energy is converted using low-head turbines and generators (see Fig. 8.11).

Fig. 8.11 Illustration of the working principle of a floating overtopping wave energy converter (taken from [14])

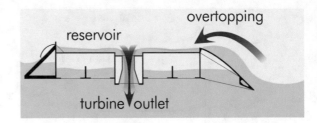

Hydro turbines have the benefit of being a mature technology that has been used for many decades for power generation. Kaplan turbines were used in the Wave Dragon device [14] and the Danish Wave Power system [15]. A Kaplan turbine is a reaction turbine that comprises a rotating element called a runner fully immersed in water, enclosed within a pressure casing (see Fig. 8.12). The turbine is equipped with adjustable (or fixed) guide vanes regulating the flow of water to the turbine runner. The blades of the runner are also adjustable from an almost flat profile for low flow conditions to a heavily pitched profile for high flow conditions.

Fig. 8.12 Schematic of a cross-section view of a Kaplan turbine (*top*) and bottom view of the runner (*bottom*) (taken from [41])

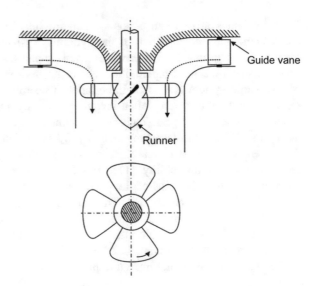

Hydro turbines can operate at efficiency values of an excess of 90 % and require low maintenance. For wave energy conversion, the bottleneck resides in the energy extraction from the waves being able to deliver sufficient head and flow for the Kaplan turbine generator unit to be economical.

8.2.5 Direct Mechanical Drive Systems

A direct mechanical drive PTO system consists on translating the mechanical energy of an oscillating body subjected to waves into electricity by means of an extra mechanical system driving a rotary electrical generator. This type of PTO system is illustrated in Fig. 8.13. For example, the mechanical conversion system can comprise gear box, pulleys and cables. Flywheel can be integrated in a rotation based system so as to accumulate or release energy and thereby smooth out power variations [16].

Fig. 8.13 Illustration of a direct mechanical drive PTO system

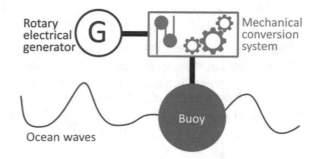

One advantage of that type of PTO system is that only up to three energy conversions are necessary, resulting in high efficiency. On the other hand, the direct mechanical drive system undergoes uncountable load cycles, and reliability of this type of system still needs to be proven.

8.2.6 Direct Electrical Drive Systems

Direct electrical drive PTO systems refer to systems for which the mechanical energy captured by the primary converter is directly coupled to the moving part of a linear electrical generator [17, 18]. Development of permanent magnets and advances in the field of power electronics have rendered this solution attractive. Figure 8.14 illustrates a direct electrical drive PTO system. A translator on which alternating polarity magnets are mounted is coupled to a buoy. The ocean waves induce a heaving motion to this system with respect to a relatively stationary stator equipped with coils, inducing electrical current in the stator.

Fig. 8.14 A schematic of a direct linear drive system for wave energy conversion (taken from [40])

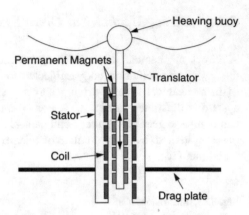

As the wave motion is directly converted to electricity in direct electrical drive PTO systems, rectification is necessary before conversion into a sinusoidal fixed voltage and frequency waveform for grid connection. This can be done either passively or actively [19]. Careful design of the mounting structure is also necessary in order to maintain fine air gaps between the translator and the stator.

8.2.7 Alternative PTO Systems

Other types of PTO system for wave energy conversion are investigated. One alternative makes use of dielectric elastomer [20, 21]. The principle is to coat with electrodes a membrane of dielectric elastomer. The mechanical energy from the waves deforms the membrane, reducing the capacitance and thereby increasing the electrical potentials of charges residing in the electrodes. Although promising simulation results have been shown, the technology is still far from mature.

8.3 Control Strategy of Power Take-Off System

8.3.1 Introduction

Ocean waves have a broad frequency band that changes with time and season, and present extreme events. On the other hand, wave energy converters are often designed with an oscillator having a narrow frequency range, i.e. their efficiency in absorbing wave energy peaks near their natural frequency (ω_0) [22]. This is represented schematically in Fig. 8.15.

Fig. 8.15 Representation of
the wave spectrum (*solid line*)
compared to the power
response of a narrow
spectrum wave absorber

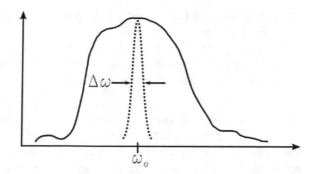

In order to render the wave energy converter more efficient by increasing the overlap between the (changing) ocean wave spectrum and the response of the converter, some tuning is necessary. The process of adapting the wave energy converter to behave as in resonance over a broad band of frequencies is referred to as *control*. The physical characteristics of the wave energy converter, like size, mass and shape, are often difficult to vary according to the incoming waves, but the behaviour of the converter can be adjusted by acting on the stiffness and/or the damping of the system.[2] These variables are accessible through the PTO system of a wave energy converter. By controlling the behaviour of wave energy converters through their PTO system, one can increase the efficiency of the system and hence its cost-effectiveness. Furthermore, in the event of extreme conditions, the wave energy converter should automatically switch to safe operation mode in order to insure its survivability. This implies a controlled system where the forces exerted on the system are monitored regularly. However necessary, control of the PTO system of wave energy converters introduces complexity to the system, which in turn lowers the reliability of the system and increases maintenance cost. The influence of the control strategy on the structural fatigue also needs to be considered [23]. Careful design of the control strategy is imperative in order to ensure cost-effective converters.

8.3.2   Types of Control Strategy

Control can be achieved on different time scales. Some of the device properties can be adjusted according to the current wave conditions, or sea state, over a period of some minutes to hours (also referred to as slow tuning). Furthermore, to allow for the irregularity of the incoming waves, the device properties should also be adapted according to the incoming wave for achieving best response, and this is referred to as fast tuning or wave-to-wave tuning.

For an unconstrained point absorber in sinusoidal wave, two conditions need to be fulfilled in order to achieve optimum control, or in other words optimum energy absorption [24]:

---

[2]For a deeper understanding of the hydrodynamics of wave energy converters, the reader is referred to Chap. 6 of this book.

(1) The velocity of the oscillator is in phase with the dynamic pressure of the incoming wave.
(2) The amplitude of the motion of the oscillator at the resonance condition needs to be adjusted so that the amplitude of the incident wave is twice the amplitude of the radiated wave from the oscillator.

The first condition corresponds to adjusting the phase of the velocity with the phase of the incoming wave and is, therefore, often referred to as phase control.

According to the second condition, the amplitude of the motion of the oscillator has to be adjusted by damping in order to achieve maximum energy conversion efficiency. If the damping is set too low, the oscillator will move too much with regards to the wave and little power will be extracted. In the same way, if the damping is too high, the amplitude of the motion will be limited, resulting in low power extraction. Hence, appropriate damping on the PTO system is fundamental.

There are many different control strategies, cf. [25]. Some of the main common ones are briefly detailed in the following.

8.3.2.1 Passive Loading Control

The damping coefficient is defined as the ratio of the force to velocity for linear motions, or torque to angular velocity for rotating motions. The damping coefficient is frequency dependent and can be either determined numerically or derived from experimental tank testing. This control strategy corresponds to adjusting the damping coefficient provided by the PTO system for a given sea state condition. For example, for rotating motion the PTO system will provide a given counter torque for a certain angular velocity of the shaft. The force-velocity (torque-angular velocity) relationship can be linear, as well as exponential or even having more advanced features (see Fig. 8.16). This technique can also be used to limit the range of movement of a device in order to avoid damaging the device in extreme wave conditions.

Fig. 8.16 Various types of linear and non-linear passive loading

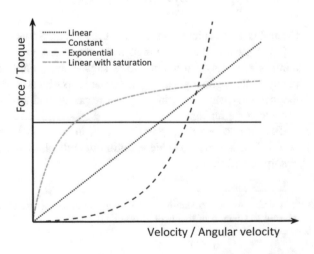

8.3.2.2   Latching Control

Latching control is a non-linear control strategy that consists of stalling the device when its velocity is zero and releasing it when the excitation force has a given phase that maximises energy absorption [26], as illustrated in Fig. 8.17. This type of control requires a PTO system that can react quickly to a given control command, like for example a hydraulic PTO system [25], and has been shown to increase significantly the absorbed energy of different devices in irregular wave conditions [27–30]. The main drawback of this strategy is that it requires the knowledge of the future wave profile in order to know when to fix and release the device, and accurate algorithms for wave prediction of wave algorithms are a challenge in itself. Latching can also lead to very large forces and it becomes less effective for two bodies system.

Fig. 8.17 Illustration of the latching control where a heaving body is kept at a fixed vertical position for a certain time interval in order to achieve phase control

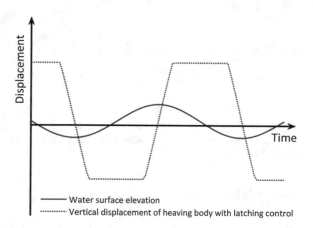

—— Water surface elevation
··········· Vertical displacement of heaving body with latching control

8.3.2.3   Reactive Loading Control

One consequence of optimum control is that some energy is returned into the sea for a small fraction of the oscillation cycle [31]; for this reason optimum control is also known as reactive control. Reactive loading control can be used to widen the frequency band of the wave energy converter around the natural frequency [25].

Any wave energy converter has inertia, which consists of the intrinsic inertia of the converter plus the inertia of the adjacent water. A wave energy converter is also often associated with a stiffness term. When pushing a body down in the water and releasing it, the body will come back to its original position after some oscillations in the same way as a mass spring system in the presence of friction would behave. Inertia is the resistance to acceleration, and stiffness is the resistance to deflection. Intuitively, those two variables should be minimised. Reactive loading control strategy aims at maximising the energy absorption at all frequencies by dynamically adjusting the spring constant (stiffness), the inertia and the damping of the oscillator.

Even though this strategy can enhance wave energy absorption [32, 33] as illustrated in Fig. 8.18, it leads to reversible and very complex PTO mechanisms. Many different suboptimal control strategies have been proposed for wave energy conversion to simplify the problem [34–36].

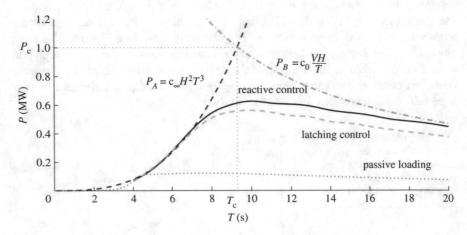

Fig. 8.18 Comparison of the power that can be absorbed from a given sinusoidal wave by a semi-submerged sphere heaving with optimum condition for three different control strategies (taken from [42])

8.4 Conclusion

This chapter introduced what a PTO system of a wave energy converter is, described the different types of PTO systems and presented the concept of control, with the overall objective of showing how crucial this subsystem is. The efficiency of the PTO system directly affects the annual energy production of the machine, and the choice of components has a direct influence on the cost of the whole converter and the maintenance cost of the system. An efficient, maintenance-free and reliable PTO system is fundamental in order to reach the goal of cost-effectiveness for wave energy conversion.

References

1. SI OCEAN: Ocean energy: cost of energy and cost reduction opportunities (2013)
2. Previsic, M, Bedard, R., Hagerman, G., Siddiqui, O.: System level design, Performance and costs for San Francisco California Pelamis offshore wave power plant, E2I EPRI Global—006A—SF report (2004)
3. Marquis, L.A.: PTO system. Presentation at a meeting of the Danish Partnership for wave energy (2014)

4. http://www.pico-owc.net/

5. Alcorn, R.G., Beattie, W.C.: Power quality assessment form the LIMPET wave-power station. In: Proceedings of the 11th International Offshore and Polar Engineering Conference (ISOPE '01), vol. 1, pp. 575–580 (2001)

6. Thielbaut, F., Sullivan, D.O., Kratch, P., Ceballos, S., Lopez, J., Boake, C, Bard, J., Brinquete, N., Varandas, J, Gato, L.M.C., Alcorn, R., Lewis, A.W.: Testing of a floating OWC device with movable guide vane impulse turbine power take-off. In: Proceedings of the 9th European Wave and Tidal Energy Conference, Southampton UK, paper no. 159 (2011)

7. Takao, M., Setoguchi, T.: Air turbines for wave energy conversion. Int. J. Rotating Mach. **2012** (2012). Article ID 717398

8. http://www.oceanlinx.com/projects/past-projects/mk1-2005

9. http://ecogeneration.com.au/news/pelamis_wave_power_powers_up_in_north_portugal/ 42829

10. Artemis intelligent power LTD. http://www.artemisip.com/

11. Ehsan, Md, Rampen, W.H.S., Salter, S.H.: Modeling of digital-displacement pump-motors and their application as hydraulic drives for nonuniform loads. J. Dyn. Syst. Meas. Contr. **122**, 210–215 (2000)

12. Henderson, R.: Design, simulation, and testing of a novel hydraulic power take-off system for the Pelamis wave energy converter. Renew. Energy **31**, 271–283 (2006)

13. Babarit, A., Guglielmi, M., Clément, A.H.: Declutching control of a wave energy converter. Ocean Eng. **36**, 1015–1024 (2009)

14. Wave Dragon, http://www.wavedragon.net/

15. Nielsen, K., Remmer, M., Beattie, W.C. 1993. Elements of large wave power plants, 1st European Wind Energy Conference (EWEC)

16. Yoshida, T., Sanada, M., Morimoto, S., Inoue, Y: Study of flywheel energy storage system for power leveling of wave power generation system. In: Proceedings of the 15th International Conference on Electrical Machines and Systems, pp. 1–5 (2010)

17. Baker, N.J., Mueller, M.A.: Direct Drive Wave Energy Converters, pp. 1–7. Power Engineering, Revue des Energies Renouvelables (2001)

18. Mueller, M.A.: Electrical generators for direct drive wave energy converters. IEE Proc. Gener. Transm. Distrib. **149**, 446–456 (2002)

19. Baker, N.J., Mueller, M.A., Brooking, P.R.M.: Electrical power conversion in direct drive wave energy converters. In: Proceedings of the European Wave Energy Conference, Cork, Ireland, pp. 197–204 (2003)

20. Moretti, G., Fontana, M., Vertechy, R.: Model-based design and optimization of a dielectric elastomer power take-off for oscillating wave surge energy converters. Meccanica **50**, 2797–2813 (2015)

21. Vertechy, R., Rosati Papini, G.P., Fontana, M.: Reduced model and application of inflating circular diaphragm dielectric elastomer generators for wave energy harvesting. J. Vibr. Acoust. **137**, 011004 (2015)

22. Falnes, J.: Ocean Waves and Oscillating Systems. Cambridge University Press (2002)

23. Ferri, F., Ambühl, S., Fischer, B., Kofoed, J.P.: Balancing power output and structural fatigue of wave energy converters by means of control strategies. Energies **7**, 2246–2273 (2014)

24. Budal, K., Falnes, J.: A resonant point absorber of ocean-wave power. Nature **256**, 478–479 (1975)

25. Salter, S.H., Taylor, J.R.M., Caldwell, N.J.: Power conversion mechanisms for wave energy. Proc. Inst. Mech. Eng. Part M: J. Eng. Marit. Environ. **216**, 1–27 (2002)

26. Budal, K., Falnes, J.: Interacting point absorbers with controlled motion. Power from Sea Waves, pp. 381–399. Academic Press, London (1980)

27. Korde, U.A.: Control system applications in wave energy conversion. In: Proceedings of the OCEANS 2000 MTS/IEEE Conference and Exhibition, Providence, Rhode Island, USA, vol. 3, pp. 1817–1824 (2000)

28. Babarit, A., Duclos, G., Clément, A.H.: Comparison of latching control strategies for a heaving wave energy in random sea. Appl. Ocean Res. **26**, 227–238 (2004)

29. Babarit, A., Clément, A.H.: Optimal latching control of a wave energy device in regular and irregular waves. Appl. Ocean Res. **28**, 77–91 (2006)
30. Falcão, A.F.O.: Phase control through load control of oscillating-body wave energy converters with hydraulic PTO system. Ocean Eng. **35**, 358–366 (2008)
31. Budal, K., Falnes, J.: Optimum operation of improved wave-power converter. Mar. Sci. Commun. **3**, 133–150 (1977)
32. Korde, U.A.: Efficient primary energy conversion in irregular waves. Ocean Eng. **26**, 625–651 (1999)
33. Valério, D., Beirão, P., Sá de Costa, J.: Optimisation of wave energy extraction with the Archimedes wave swing. Ocean Eng. **34**, 2330–2344 (2007)
34. Price, A.A.E.: New perspectives on wave energy converter control. Ph.D. thesis. University of Edinburgh (2009)
35. Hals, J.: Modelling and phase control of wave-energy converters. Ph.D. thesis. Norwegian University of Science and Technology (2010)
36. Hansen, R.H.: Design and control of the power take-off system for a wave energy converter with multiple absorbers. Ph.D. thesis. Aalborg University (2013)
37. Raghunathan, S.: The Wells air turbine for wave energy conversion. Prog. Aerosp. Sci. **31**, 335–386 (1995)
38. Setoguchi, T., Santhakumar, S., Maeda, H., Takao, M., Kaneko, K.: A review of impulse turbines for wave energy conversion. Renew. Energy **23**, 261–292 (2001)
39. Finnigan, T., Auld, D.: Model testing of a variable-pitch aerodynamic turbine. In: Proceedings of the 13th (2003) International Offshore and Polar Engineering Conference, Honolulu, Hawai, USA, May 25–30 2003
40. Drew, B., Plummer, A.R., Sahinkaya, M.N.: A review of wave energy converter technology. Proc. Inst. Mech. Eng. Part A: J. Power Energy **223**, 887–902 (2009)
41. IIT. 2016. Basic principles of turbomachines. http://nptel.ac.in/courses/112104117/chapter_7/7_11.html. Accessed 19 Jan 2016
42. Falnes, J., Hals, J.: Heaving buoys, point absorbers and arrays. Philos. Trans. R. Soc. A **370**, 246–277 (2012)

Chapter 9
Experimental Testing and Evaluation of WECs

Arthur Pecher

9.1 Overview

The main objective of a test campaign is to investigate some aspects of a technology or to validate them. Experimental test campaigns on a (model or subsystem) of a WEC can be done in three different environments: in a controlled and wet environment (referred to by tank testing), a controlled and dry environment (referred to by test bench) and in an uncontrollable wet environment (referred to by sea trials). Experimental tests on the full or subsystems of the device can be performed during all the development stages of the WEC, and well beyond (Fig. 9.1).

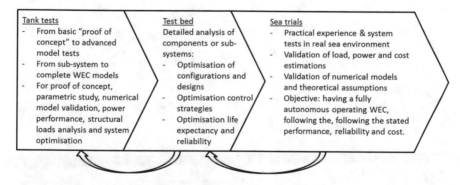

Fig. 9.1 Flow chart of the experimental testing of WECs

A. Pecher (✉)
Department of Civil Engineering, Aalborg University,
Thomas Manns Vej 23, 9220 Aalborg Ø, Denmark
e-mail: apecher@gmail.com

© The Author(s) 2017
A. Pecher and J.P. Kofoed (eds.), *Handbook of Ocean Wave Energy*,
Ocean Engineering & Oceanography 7, DOI 10.1007/978-3-319-39889-1_9

As the complexity, time and costs of the tests significantly increase with the size of the model and the complexity of the environment, it is (often) more adequate to investigate system variables on smaller models, even if the necessity of investigating these parameters come at a later technological readiness level (TRL) stage of the development process (Chap. 4 and [1]). That is why during the development of WECs, there will be a continuous change in between the size of the model and the test environment.

In Fig. 9.2, some of the main laboratory models, test benches and prototypes that have so far been used during the development of the Wavestar WEC are presented. This just illustrates that several different experimental test campaigns are required to support the numerical work and the general development of a WEC.

Fig. 9.2 Some of the development efforts behind the Wavestar WEC; *top row* an early stage lab model (TRL 2), a benign site prototype (TRL 5); *second row* a sea trial prototype (TRL 7) and a hydraulic PTO test bench; *bottom row* a large single float lab model and a lab model with advanced PTOs. Courtesy of Wavestar

9.2 Tank Testing

9.2.1 Overview

Tank tests can be performed during the whole development process of a WEC, from its early proof of concept up to preparing for the next serial production unit. The objectives and testing procedures of a tank testing campaign can thereby differ, depending on the project. Before knowing an exact location of installation of a commercial device, all the tests should be performed on a generic basis, while afterwards more precise investigations with the exact environmental conditions can be done. Table 9.1 presents an overview of possible tank test campaigns together with their main objectives.

Table 9.1 Overview of possible objectives for performing tank tests

Objectives	Sea states	Objective	Comment
Proof of concept	Operational	Verify if the model produces useable energy	Measurements of the produced power is required
Power performance evaluation	Operational	Power performance curve or surface, enabling the estimation of the mean annual energy production at one or multiple locations	The PTO load needs to be optimised separately for each sea state
Assess structural and mooring forces	Design (operational)	Indication of maximum loads and effectiveness of survival mechanism	Influence of some additional environmental parameters need to be addressed as well
Parametric study	Operational and design	Assessing the influence of environmental and design parameters on: – power performance – motions and loads – seakeeping or mooring forces	Results have to be compared to the reference setup. One parameter has to be assessed at the time
PTO control	Operational	– Improved power performance – Reduced loads or improved lifetime of components	
Hydrodynamic response	Operational and design	– Natural periods of oscillations – Response amplitude operators (RAO) – Effect of mooring on the motions	Can be done for the whole structure and all moving wave-interacting bodies
Evaluate the numerical models	Operational and design	Validation, calibration and evaluation of numerical models	Numerical models can be very helpful in the development of WEC

Note that all the tank tests related to power performance and structural and mooring loads need to be performed in irregular waves. Regular waves should only be used to characterise the device or to calibrate the system or numerical model (see Chap. 10), as it does not represent a realistic marine environment. In order to have a decent statistical ground, the duration of a test in irregular waves should include a minimum of 1000 waves (of the predominant wave period T_p).

The environmental parameters that can have an influence on the power performance and on the maximal structural and mooring forces can be of many kinds, for example (non-exhaustive) (see Chap. 3):

- Water depth, as it has a strong influence on the wave steepens, wave celerity and wave direction
- The wave spectrum, which can be composed of different wave components coming from different storms (this defines the content of the irregular waves)
- Directional spreading (defines the direction related to the wave spectrum)
- Water currents (which can result e.g. from the wind, tides or near-shore effects and can have an influence on the motion, directionality and loads on the WEC)
- Wind (can have an influence on the motion, directionality and loads on the WEC)

Besides the environmental parameters, other parameters might have an influence on the wave conditions such as the disposition of the wave energy converters array.

9.2.2 Representative Sea States

9.2.2.1 Operational Sea States

A description of the wave conditions at a certain location may be required in even a more condensed way than given by a scatter diagram. This is often the case for tank testing, as it would be too time-consuming to assess and optimise the performance of a device for all the bins of the scatter diagram. The gain of time by reducing the to-be-tested wave conditions, will benefit the assessment and optimisation possibilities. In practice, this can be done by grouping various bins of a scatter diagram into a limited amount of "zones", also referred to as "sea states". Each sea state will then at least be characterised by a wave period and a significant wave height, and a common water depth and wave spectrum type will be used. The influences of additional environment parameters will need to be investigated separately. An example of commonly used sea states are the five operational sea states and the three design sea states representing the Danish part of the North Sea (Point 3) [2, 3].

Other examples in which sea states have been used for the estimation of the *AEP* can be found in [4–7].

The selection of sea states for the estimation of the AEP of a WEC has to be done carefully. The large variability in wave conditions between different locations (as illustrated in [8] and in Chap. 3) can result in a loss in accuracy relative to the use of the complete scatter diagram for the estimation of the *AEP*, however this can be limited if the sea states are selected carefully.

The following recommendations are for the selection and definition of the *operational* sea states:

- The amount of sea states should be limited (less than 10 preferably)
- They should be selected in order to cover the wave energy contribution diagram as well as possible, rather than the scatter diagram.
- The wave energy contribution of each sea state should be between 5 and 25 % of the total, while having a probability of occurrence of at least 0.5 % of the time, corresponding to 44 h annually.
- The same size of zones (identical intervals of H_s and T_e) can be used for the different sea states, but they can be reduced for zones with higher contribution values in order to increase their accuracy.
- As the optimal size of a WEC in terms of *AEP* (usually) increases proportionally with the wave power level of a site, it can be reasonable to have larger sizes (larger intervals of H_s and T_e) of sea states when describing more wave energetic locations.
- For the estimation of the *AEP*, there is no need to include the very small or large wave conditions, as they will not contribute significantly to the *AEP* [6]. This is due to their low wave energy contribution and a WEC has usually a bad performance in them, as their design is normally not optimized for them [5].

Note that the bins that are not included into sea states will not be accounted for in the AEP estimation.

The sea state selection in Fig. 9.3 contains seven sea states with the same parameter intervals of 2 m and 2 s for the respective H_s and T_z axis. They represent 90 % of the wave energy resource and 89 % of the probability of occurrence. In other words, 7801 out of the 8766 annual hours are included and an average wave power level over the included bins is of 26.3 kW/m instead of the 29.3 kW/m, which can be derived from the whole scatter diagram. These values could be increased by adding more sea states, however (as previously mentioned) the largest loss in *Prob* and *Contrib* is in the smallest (below 0.5 m H_s) and largest (above 8.5 m H_s), which are not important as mentioned before.

Scatter diagram

Hs \ Tz	3.5	4.5	5.5	6.5	7.5	8.5	9.5	10.5	11.5	12.5	13.5	14.5
0.25	0.0066	0.0056	0.0030	0.0023	0.0011	0.0007	0.0003	0.00005				
1	0.0453	0.1650	0.0906	0.0347	0.0131	0.0047	0.0019	0.00069	0.0001	0.00004	0.00007	0.00005
2	0.0018	0.0368	0.1604	0.0650	0.0229	0.0099	0.0032	0.00121	0.00009	0.00005	0.00005	
3		0.0003	0.0187	0.1084	0.0335	0.0071	0.0033	0.00171	0.0004	0.00007		0.00002
4			0	0.01021	0.05565	0.01163	0.00209	0.00052	0.00034	0.00021	0.00005	
5				0.00002	0.00729	0.02391	0.00301	0.00069	0.00031	0.00014	0.00005	0.00005
6					0.00012	0.00603	0.00691	0.00052	0.00007			
7				0.00002	0.00009	0.00026	0.00352	0.00152	0.00016	0.00005		
8							0.00062	0.00288	0.00017			
9								0.00086	0.00073	0.00002		
10								0.00002	0.00043	0.00016		
11									0.00011	0.00014		
12										0.00004		

Wave Energy Contribution

Hs \ Tz	3.5	4.5	5.5	6.5	7.5	8.5	9.5	10.5	11.5	12.5	13.5	14.5
0.25	0.00003	0.00003	0.00002	0.00002	0.00001	0.00001						
1	0.0032	0.015	0.010	0.0046	0.0020	0.00082	0.00036	0.00015	0.00002	0.00001	0.00002	0.00001
2	0.0005	0.014	0.072	0.034	0.014	0.0068	0.0025	0.0010	0.00008	0.00005	0.00006	
3		0.00021	0.019	0.129	0.046	0.011	0.0057	0.0033	0.0008	0.0002		0.00005
4			0	0.022	0.14	0.032	0.0065	0.0018	0.0013	0.00086	0.00022	
5				0.00007	0.028	0.10	0.015	0.0037	0.0018	0.0009	0.0003	0.0004
6					0.0007	0.038	0.048	0.0040	0.0006			
7				0.0001	0.0007	0.0022	0.033	0.016	0.0018	0.0006		
8							0.0077	0.039	0.0026			
9								0.015	0.014	0.0004		
10									0.010	0.004		
11									0.003	0.004		
12										0.001		

EMEC - Billia Croo

Sea State	Hs [m]	Tz [s]	Te [s]	Contrib [-]	Prob [-]	Pwave [kW/m]	Pwave*Prob [kW/m]
1	1.52	5.2	6.4	0.11	0.45	7.2	3.24
2	1.72	6.8	8.3	0.06	0.14	11.9	1.61
3	3.09	6.4	7.8	0.17	0.14	36.3	4.97
4	3.66	7.7	9.4	0.23	0.11	61.4	6.61
5	5.18	8.3	10.1	0.17	0.04	133.4	4.97
6	5.69	9.6	11.7	0.07	0.01	186	2.06
7	7.43	10.1	12.3	0.10	0.01	332	2.83
			sum	0.90	0.89		26.3

Fig. 9.3 Example of a possible sea states selection for Billia Croo at EMEC, which are represented on the scatter diagram (*top figure*) and on the wave contribution diagram (*middle*) and summarized in the table

The wave energy contribution of every bin of the scatter diagram to the overall wave energy resource can be calculated by:

$$Contrib_{bin} = \frac{(P_{wave})_{bin} \cdot Prob_{bin}}{\sum_{bin=1}^{n}((P_{wave})_{bin} \cdot Prob_{bin})} \tag{9.1}$$

The characterizing H_s and T_e values of every sea state are the average of the environmental parameters of the various bins included in a sea state weighted by their corresponding probability of occurrence. Herewith, the corresponding wave power (P_{wave}), which should take the water depth into account, can be calculated by

$$P_{wave} = \frac{\rho g^2}{64\,\pi} H_{m0}^2 T_e \left[1 + \frac{2\,k_e h}{\sinh 2\,k_e h}\right] \tanh k_e h \qquad (9.2)$$

(More details in Chap. 3). The corresponding equations to calculate the characterizing H_s and T_e for each sea state are:

$$H_{s\,SS} = \sqrt{\frac{\sum_{SS,\,bin=1}^{n} H_{s_{SS,\,bin}}^2 \cdot Prob_{SS,\,bin}}{\sum_{SS,\,bin=1}^{n} Prob_{SS,\,bin}}} \qquad (9.3)$$

and

$$T_{e_{SS}} = \frac{\sum_{SS,\,bin=1}^{n} T_{e\,SS,\,bin} \cdot Prob_{SS,\,bin}}{\sum_{SS,\,bin=1}^{n} Prob_{SS,\,bin}} \qquad (9.4)$$

While, the probability of occurrence ($Prob_{SS}$) and the wave energy contribution ($Contrib_{SS}$) of a sea state correspond to the sums of the respective values of the bins that each of them include.

Further information and more advanced approaches to representation of the wave climate can be found in Kofoed and Folley [9]

9.2.2.2   Design Sea States

The design sea states correspond to a set of wave conditions with large wave heights, in which normally the largest loads on the structure and mooring system are expected. These often correspond to the 50 or 100 year return wave height (depending on the design standard e.g. [10]), and related wave period. These design sea states can be obtained by defining certain return periods of these extreme wave events and can be derived from long-term probability distributions that are based on past events and hindcast data (Fig. 9.4).

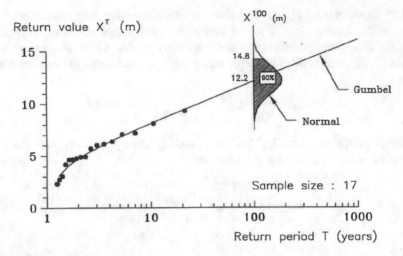

Fig. 9.4 Illustration of the possible process to define the design wave height for a certain return period, which in this case is 12.2 for a return period of 100 years without considering sample variability or 14.8 m with sample variability at a 90 % one-sided confidence interval [11]

The design wave height with a certain return period can be obtained through those long term probability distribution, while there is no theory to determine the corresponding wave period, due to the complexity and locality of the joint distribution between wave height and wave period (an example of this is a scatter diagram). The design sea state conditions are chosen corresponding to the design wave height and a range of possible corresponding wave periods [10–12].

In the following practical example, the corresponding wave heights to the different return periods have been obtained through a Peak-over-Threshold analysis, using a Generalized Pareto distribution, based on the 30 years hindcast data [13, 14]. The related wave periods to the design wave heights have been obtained through fitting them on a trendline going through all the data points (Fig. 9.5).

It is of importance to investigate other environmental parameters and site conditions that could influence the loads on the WECs. Most of these environmental parameters are the same that need to be investigated for power production, however some others need to be addressed as well [10, 12, 15]:

- Breaking waves
- Ice
- Current
- Water level variations

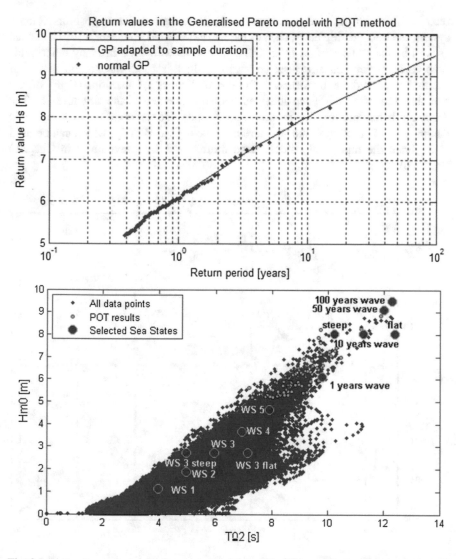

Fig. 9.5 Estimation of the design wave heights for different return periods (*top*) and representation of the operational and design sea states superposed on the 30 year hindcast scatter diagram of DanWEC Pt 1 (*bottom*)

9.2.3   Hydrodynamic Response

9.2.3.1   Natural Period

The natural period of oscillation, also referred to as the frequency of free oscillation or the eigen period, reveals the decaying period at which a mechanical system recovers from an initial displacement until its undisturbed rest position. Such a system can be a floating body, such as the wave-activated bodies of a WEC, but can as well be the water surface in an OWC or it is as well used in many other fields such as acoustics and structural engineering. In this case, the natural period of oscillation should be investigated for the 6 degrees of freedom (DoF) of the structure and of all the wave-activated bodies connected to the structure [2]. The decaying behaviour of the motion also reveals the amount of hydrodynamic damping that is present in the structure or wave-activated body, which can be used to calibrate the numerical models (read more about hydrodynamics in Chap. 6) (Fig. 9.6).

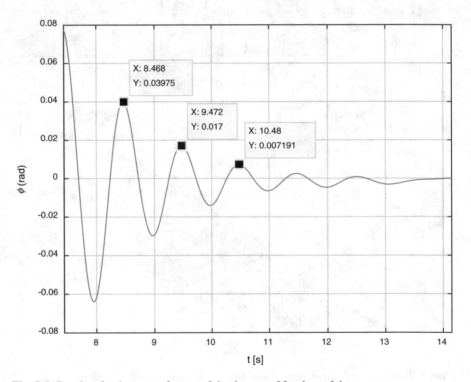

Fig. 9.6 Results of a decay test for one of the degrees of freedom of the system

The damping in the oscillations of a floating body arise from the wave-drift damping of the hull (wave radiation and diffraction), the viscous nature of water and possibly from the influence of other systems such as the mooring or PTO (more details in Chap. 6). The damping ratio ζ derived from the logarithmic decrement δ can further describe the decaying motion besides the natural period of oscillation. These can be calculated by the following equations:

$$\delta = \frac{1}{n} \ln \frac{x_0}{x_n} \tag{9.5}$$

$$\zeta = \frac{1}{\sqrt{1 + \left(\frac{2\pi}{\delta}\right)^2}} \tag{9.6}$$

where

- x_0 is the value of the first amplitude
- x_n is the value of the amplitude of a peak n oscillations (periods) away

In practice, these tests assessing the natural period of oscillations for a degree of freedom (DoF) can be performed by applying a force on the structure that would force the body to oscillate only in the specific degree of freedom under investigation. The effect of possible coupled secondary systems, such as mooring system, can be evaluated by repeating the test with and without them.

The natural period of oscillation of a body in any degree of freedom (T_x) is dependent on its mass (m) and its geometry. The geometry affects its (hydrodynamic) added mass $M_{a,0}$ and thereby the stiffness of the system k. Depending on the DoF, these can as well be expressed in terms of mass moment of inertia I_{yy} and added mass moment of inertia I_{yy}^a. The stiffness in heave is in function of the density of the fluid and the waterplane area A_{wp}, while for pitch and roll it depends on the metacentric height $\overline{GM}$ [16].

Generic resonance period equation:

$$T_x = 2\pi \sqrt{\frac{m + M_{a,0}}{k}} \tag{9.7}$$

Heave resonance period:

$$T_H = 2\pi \sqrt{\frac{m + M_{a,0}}{\rho g A_{wp}}} \tag{9.8}$$

Pitch resonance period:

$$T_{pitch} = 2\pi \sqrt{\frac{I_{yy} + I_{yy}^a}{\Delta GM_g}} \tag{9.9}$$

Roll resonance period:

$$T_{roll} = 2\pi \sqrt{\frac{I_{xx} + I_{xx}^a}{\Delta GM_g}} \tag{9.10}$$

In order to modify the resonance frequency, the size, the mass, the shape or the inertia of the body can be changed.

9.2.3.2 Response Amplitude Operators

The response motion of a wave-activated body under wave interaction (corresponding to a forced excitation) can be assessed in regular and irregular waves. The analysis in regular waves presents a direct visual and intuitive representation of the response motions. The response amplitude operators (RAOs) are the ratio between the amplitudes of the motion in one of the degrees of freedom and the amplitudes of the incoming the waves. The RAOs can also be derived from the motion and wave spectra, as the spectra are proportional to the amplitudes squared. The ratios between the spectra of the motion and the spectra of the incoming waves are denote the transfer function, which therefore is the RAO squared.

In Fig. 9.7, the response motion in one degree of freedom of a wave activated body is represented. The successive tests were done with equal wave height but incrementing wave periods. Each individual test should last for 30–120 s, which should be sufficient to have a stable motion.

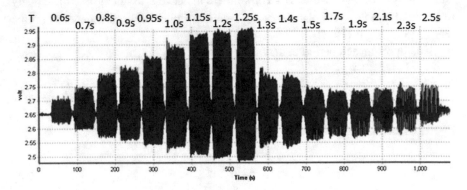

Fig. 9.7 Illustration of the response motions in 1 DoF of a wave-activated body in regular waves having a same wave height but incrementing wave period (The *vertical axis* represents the absolute motion of the body)

The wave period presenting the largest motions, correspond to the resonance period of the wave activated body. The regular wave trials could also give information concerning the phase difference between the resulting motion of the device and the excitation. This can in some cases be a very interesting feature as at a phase shift of 90° resonance occurs.

Figure 9.8 illustrates the transfer function of the pitch motion of a device obtained experimentally in regular and irregular waves. **Note** that (in this particular case) the wave spectrum drops to zero around the resonance frequency. This makes it difficult to obtain a reliable transfer function for these frequencies and therefore other tests should be performed that covers better the corresponding wave periods.

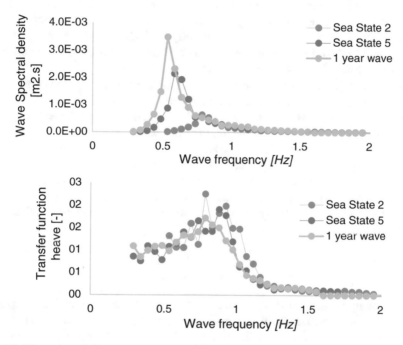

Fig. 9.8 Illustration of the wave spectrum and in heave obtained in different irregular waves

Note that structures such as ships or floating oil platforms, that are required to be stable at all times, will be designed so that their resonance frequencies will be outside the range of the wave spectrum. This is not always true for WECS, as they might want the main reference structure to be stable, while their wave absorbing body (/-ies) could benefit from the larger resonance oscillations, e.g. the Weptos WEC or the Wavestar WEC.

9.2.4 Power Performance Evaluation

9.2.4.1 Introduction

For experimental tank testing, it is suggested to measure the absorbed power P_{abs} by the device as early as possible in the power conversion chain (from wave-to-wire), in order to obtain the best representation of its wave power capturing abilities. The following power conversion stages should not be incorporated in the performance measurement, as they physically could be subject to serious modifications and their losses and design are usually difficult to scale. This P_{abs} by the model can be converted using different kind of PTOs, e.g. mechanical, pneumatic or hydraulic, which depends on the working principle of the device.

Based on P_{abs} and the available wave power to the device P_{wave}, the performance can be expressed by a non-dimensional performance ratio called the "capture width ratio" or CWR (or η). The fact that it is non-dimensional presents the advantage that the same results can be used for different scaling ratios of the device, meaning that only the wave parameters needs to be adapted correspondingly to the scaling ratio and site. The available wave power to the device corresponds to the average wave energy content per meter of wave front multiplied by the characteristic or active width of the device (width$_{active}$), which corresponds to the width of all the components of the device that are actively involved in the primary conversion stage from wave to absorbed energy.

$$CWR = \frac{P_{abs}}{P_{wave} \cdot width_{active}} \tag{9.11}$$

Depending on the tested wave conditions, a performance curve (2-dimensional) or surface (3-dimensional) can be created, which represents CWR relative to one or two wave parameters and are illustrated in Fig. 9.9. The performance curve or surface could present CWR relative to its most influential wave parameter (T_p or H_s), which should be determined during the tank tests, or relative to corresponding non-dimensional values. T_p could for example be made non-dimensional by dividing its corresponding wavelength in deep water ($L_{p,0}$) by the diameter of the main wave absorbing body (d), while H_s could directly be divided by d to obtain a dimensionless parameter [17]. Note that the most influential parameter can in this case easily be derived from the performance surface, as CWR increases significantly with decreasing T_p, while it remains relative constant for different values of H_s.

Note that the peak wave period (T_p) is used as reference in Fig. 9.9, instead of T_e, as the wave frequency spectrum for all the lab tests was user-defined. During sea trials, the shape of the wave frequency spectrum changes with the conditions; there it is more representative to present the performance relative to T_e, as it presents a more robust average of the wave conditions. T_p represents only one parameter of the spectrum (the peak), which makes it very unstable and not very representative. T_e is derived from the whole spectrum, which is thereby more reliable.

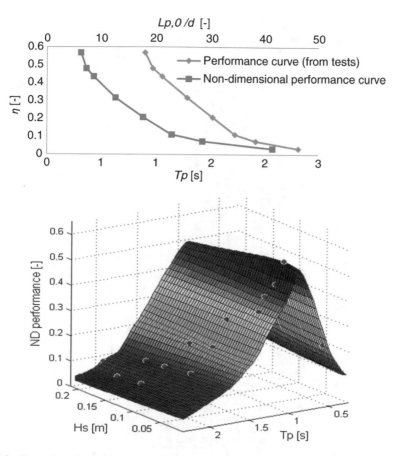

Fig. 9.9 Illustration of a performance (in terms of $CWR = \eta = ND$ performance) curve and surface that can be obtained through tank testing [18]

9.2.4.2 Power Performance Estimation Based on Sea States

Based on the performance of the device obtained in irregular waves (the influence of 3D waves will be assessed through a parametric study) and after optimization of the PTO load in the tested sea states, the mean annual energy production (*MAEP*) of the device can be estimated. Table 9.2 illustrates how the performance of a device, based on tank testing with sea states, is calculated and can be presented.

The upper part summarizes the wave characteristics and corresponding performance results for the full-scale model based on *CWR* obtained through tank tests. The bottom part presents the resulting values that give an overall overview of the performance. Definitions of the different terms are given after the table. An estimation of the actual generated electrical power (P_{el}) can also be added if the efficiency of the PTO system (η_{PTO}) of the full-scale device is known. This value can (normally) not be deducted from the tank tests as the rest of the conversion

Table 9.2 Illustration of the performance results and the *AEP* estimation based on tank testing with sea states

	Sea state	H_s (m)	T_e (s)	P_{wave} (kW/m)	Prob (–)	Contrib (–)	η (–)	P_{abs} (kW)	$Prob * P_{abs}$ (kW)	η_{PTO} (–)	P_{el} (kW)
Sea state	1	1	4.8	2.4	0.468	0.07	0.32	92	43	0.88	81
	2	2	6	11.8	0.226	0.16	0.37	524	118	0.9	472
	3	3	7.2	31.7	0.108	0.21	0.25	951	103	0.92	875
	4	4	8.4	65.8	0.051	0.21	0.14	1105	56	0.9	995
	5	5	9.6	117.6	0.024	0.17	0.08	1129	27	0.88	993
Overall	Sum				0.88	0.82			348		
	Weighted average			16.3			0.19	348		0.89	310
	Maximum P_{abs} (kW)							1129			1005
	MAEP (MWh/year)							3048			2713
	Capacity factor (–)							0.31			

chain after the energy absorption, from wave-to-wire, will normally not be included in the model or not be representative of its full-scale version in terms of efficiency.

The non-dimensional performance or *CWR* of the device for each sea state (η_{SS}), as described in Sect. 9.2.4.1 and by Eq. (9.11), is the ratio between the absorbed power by the device and the wave power available to the device for the wave conditions corresponding to the respective sea state.

The average absorbed power by the device or P_{abs} for a sea state can be calculated with the η_{SS} and P_w for the wave conditions corresponding to a sea state:

$$(P_{abs})_{SS} = \eta_{SS} \cdot (P_w)_{SS} \cdot width_{active} \tag{9.12}$$

The overall non-dimensional performance can be obtained by the weighted sum of the η_{SS} relative to their wave energy contribution:

$$\eta_{overall} = \sum_{WS=1}^{n} \eta_{SS} \cdot (Contrib)_{SS} \tag{9.13}$$

The overall average absorbed power $P_{overall}$ is calculated by the weighted sum of the P_{abs} of each sea state relative to their *Prob* or by taking the product of the overall non-dimensional performance and the available wave power. Note that P_{wave} corresponds to the overall available wave power calculated based on the scatter diagram (the gross available or theoretical wave resource) and not only what is included in the sea states. If the calculation of $\eta_{overall}$ is based on the technical resource, then this will be significantly overestimated.

$$(P_{abs})_{overall} = \sum_{SS=1}^{n} (P_{abs})_{SS} \cdot (Prob)_{SS} \tag{9.14}$$

or

$$(P_{abs})_{overall} = \eta_{overall} \cdot (P_{wave})_{Overall} \cdot width_{active} \tag{9.15}$$

The annual (absorbed) energy production (*AEP* and given in kWh) is then obtained by multiplying $(P_{abs})_{overall}$ by the duration of a year (8766 h):

$$AEP = 8766 \cdot (P_{abs})_{overall} \tag{9.16}$$

The capacity factor (*CF*) represents the average usage of the installed capacity, which corresponds to the ratio between the overall average absorbed energy and maximum absorbed energy in any wave condition. (This is based on the average absorbed power in the maximum sea state and does not take the possible maximum value of the absorbed power in the maximum sea state into account.)

$$CF = \frac{\left(P_{abs}\right)_{overall}}{maximum\left(P_{abs}\right)_{SS}} \tag{9.17}$$

Figure 9.10 presents an example of the evolution of η_{SS} over the various sea states together with their corresponding wave energy contribution, absorbed power and the product of the absorbed power with the probability of occurrence.

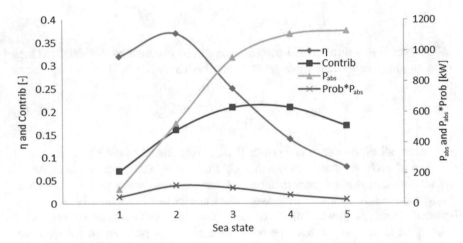

Fig. 9.10 Example of the representation of η_{SS} (*blue line*), $Contrib_{SS}$ (*red line*), P_{abs} and $Prob * P_{abs}$ for the different sea states, based on the values given in Table 2

The representation of the performance enables to visualise the various parameters over the different sea states. In this case, the non-dimensional performance peaks at sea state two while the maximum wave energy contribution peaks between sea state three and four. The mean annual energy production (MAEP) could be increased by trying to match these peaks better. This could be done by increasing the size (scaling ratio) of the full-scale device (which has been discussed in Sect. 9.2.5.3). However, P_{abs} is in the same range in sea state three, four and five. This is an advantage as the device will require roughly the same capacity of PTO system for these wave states (leading to a high capacity factor), which is most-likely not the case if the peak of the non-dimensional performance is close or beyond the peak of the wave energy contribution curve. The Prob * P_{abs} curve shows that most of the energy will in average be absorbed in sea state 2, which correspond to 1 m (H_s) waves, and the least in sea state five.

9.2.4.3 Power Performance Estimation Based on the Scatter Diagram

The performance analysis and estimation of the MAEP presented as a non-dimensional performance surface is similar to the one based on sea states. The main difference is that the performance is not only given for a limited amount of sea states, but is given individually for every "bin" of the scatter diagram. Therefore, in the equations given in Sect. 9.2.4.2, the subscript "SS" should be replaced by "bin" and they are then applicable in this context.

The power performance of the device is represented by the 3-dimensional performance surface (as illustrated in Fig. 9.9), while the wave conditions at the location can be represented by the scatter diagram and by the wave energy contribution diagram (see Sect. 9.2.2.1). The resulting performance of the full-scale device corresponds to the power matrix and P_{mech} * Prob graph, in which P_{mech} represent the mechanical absorbed power, as given in Fig. 9.11.

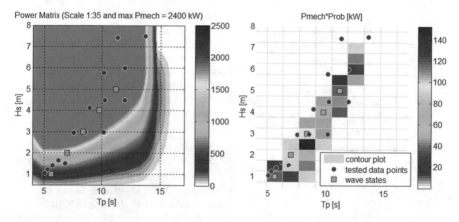

Fig. 9.11 Power matrix and P_{mech} * Prob plot with scaled tested wave conditions (*blue dots*) and corresponding sea states (*green squares*) (Illustration of the WEPTOS WEC in Danish North Sea wave conditions at a scaling ratio of 1:35 and with a maximum full-scale P_{mech} of 2400 kW)

Note that it is off course of importance that the wave energy contribution diagram is well covered by the tested sea states, otherwise excessive extrapolation will have to be done in order to obtain a representative performance curve. This is particularly difficult when a large range of scaling ratios are used.

Note also that the power matrix derived from testing can only be applied to offshore locations that have a similar environment (e.g. water depth, …) and wave conditions (e.g. spectral shape, …). To broaden the usability of this power matrix, additional parametric studies can be performed in order to investigate the influence of certain important parameters (e.g. water depth).

9.2.4.4 Testing Procedure for Power Performance Analysis

As tank testing is relatively time-consuming (from a couple of weeks to months usually) and the time in the tank facilities is limited, it is important to have a well-defined test procedure. The main steps during the power performance assessment of a WEC are:

What	How
Part 1: Identification and characterisation of the system	
Identify hydrostatic and hydrodynamic response of the WEC	– Stability and GM centre of mass and buoyance etc – Asses the natural period of oscillations and decay response – Asses the hydrodynamic response in regular waves
Identify the sensitivity and range of the PTO and sensors	– In regular waves asses the required range of the PTO and sensors, and adapt them accordingly to the maximum
Identify the influence of physical parameters, which are intended to be assessed in depth in irregular waves later	Perform elementary regular wave batches (0.5–2 min tests with increasing wave period), while changing one variable at the time between batches. This could give a good feeling on the importance of the parameter
⇒ Possibly make adjustments on design, sensors and/or scaling ratio, as now the natural period of oscillation and the range of usage of the sensors and PTO have been identified	
Part 2: actual test campaign	
Irregular waves on reference setup	– Optimise PTO load for each sea state – Duration of tests is 1000 waves (relative to Tp)
Optimisation of the design	– Asses the influence of alterations to the design, in order to optimise its power performance, hydrodynamic behaviour, …. Start with tests in sea state 2 and 3, as these contribute the most to the *MAEP* – The PTO load needs to be optimised for each sea state – These tests will also provide the final RAO's
Asses the influence of additional physical parameters	– e.g. mooring configuration, water depth, oblique waves … – The PTO load needs to be optimised for each sea state
Asses the influence of additional environmental parameters	– This is of importance when later trying to estimate the performance of the device for different wave conditions, e.g. water depth, oblique waves, 3D waves, … – The PTO load needs to be optimised for each sea state
⇒ The data of each test should be processed after each individual test in order to be able to compare the results with the one of previous tests	

To begin with, a general appreciation of the **hydrodynamic behaviour** of the device should be made. This can in practice be done in regular waves and without any PTO loading, by making various short tests (0.5–2 min each) where the wave height is maintained constant and the wave periods are each time incremented. This should be repeated for constant wave periods and increasing wave heights, and it could for example be used to identify the resonance frequency of the structure or of the wave activated body and show the range of effect of the wave conditions. A similar approach could possibly be used to investigate the influence of different configurations, for example if the device has an adaptable geometry, weight or floating level.

After the hydrodynamic behaviour, the **sensitivity and relevant working range of the PTO loading adjustment system** have to be assessed. In this case, the load should be increased again in batches (0.5–2 min each) for a couple of the tested wave conditions. In practice, this can be done by incrementing the load between each batch by 10 % of its full range and repeated for the smallest, one or two medium and the largest sea states. Although these tests are not crucial, they often lead to a significant gain in time.

Note that in order to maintain the same wave energy content in between regular and irregular waves, the significant wave height (H_{m0}) from the irregular waves has to be divided by $\sqrt{2}$ to obtain the wave height for the regular waves, while maintaining the same wave period ($T = T_e$). However, in the case that the response or performance of the device is mostly dependent on the wave period, it might be beneficial to match the wave period in regular waves with T_p, as this is the dominant wave period in irregular waves.

The actual **performance assessment** is based on long-crested irregular waves, having a specific wave spectral shape (e.g. JONSWAP spectrum with $\gamma = 3.3$). Each individual IW lab test should have a length of 1000 peak wave periods (for statistical robustness), which should take about 20–30 min, depending on the scale. Moreover, in each wave condition the PTO load needs to be optimised for optimal energy production (as presented in Fig. 9.12). Ideally, an exact reproduction of the waves should be performed in between those tests. Depending on the complexity

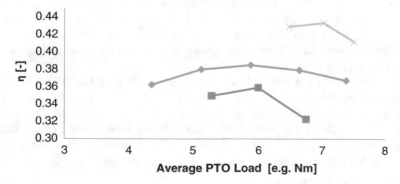

Fig. 9.12 Illustration of the effect of the PTO loading on the non-dimensional performance (η), given for three different configurations of the same WEC model

and possibilities of tuning the PTO loading, this step will require generally between 3 to 5 tests for each sea state. However, if the device contains various design variables, then they need to be investigated as well in some of the sea states.

After having obtained the best performance of the device in all the sea states, the **influence of other environmental parameters** can be investigated and the influence of some extraordinary components or modifications on the performance and hydrodynamic behaviour can be assessed. Regarding the wave conditions, the sensitivity of the performance to the wave frequency spectrum, wave direction and wave directional spreading (3D waves) should be investigated. This should be done with various values for them, probably focusing on the most wave energy contributing sea states. The load optimisation should be done for each case and their result should be compared to the reference long-crested irregular waves. The same goes for the tests analysing the influence of some extraordinary components or modifications on the performance and hydrodynamic behaviour.

9.2.5 Scaling

9.2.5.1 Defining the Scaling Ratio

The scaling ratio indicates the ratio between the model or prototype and the commercial WEC. The size of the model should be chosen in function of the laboratory facilities and the purpose of the tests, while the size of the commercial WEC depends on the WEC technology and on the commercialisation strategy of the developer, which is often a trade-off between financial resources and optimising its cost of energy. The scaling ratio, and thereby the size of the commercial WEC, can be optimised all along the development of the WEC as it has a strong influence on the overall cost and power production of the device, but as well on the capacity factor and fluctuations in the power produced by the WEC. However, in order to obtain representative wave conditions for the tank tests model, a scaling ratio needs to be used to scale the sea states and water depth.

In most cases, the first serial production of a WEC, will be smaller than the scaling ratio leading to maximum power production, in order to keep capital expenditures lower. In this case, the scaling ratio will need to be as large as the financial resources allow for it.

Whenever, the power production needs to be maximised, the scaling ratio will intend to have the resonance period of the wave-activated bodies (e.g. point absorbers, OWCs and pitching flaps) in function of the predominant wave period, which corresponds to the peak of the wave energy contribution. Other WECs, where the structure is required to be stable (such as floating overtopping devices) will try to keep their resonance period as far out as possible from the wave peak period.

In Fig. 9.13, an illustration is given on how the main influential power performance factors might overlap. Note that the resonance period of the wave absorbing body corresponds to the peak of the capture width ratio, and that the predominant wave period, corresponds to the peak of the wave energy contribution.

Fig. 9.13 Illustration of a possible overlap between the resonance period of the wave absorbing body, with the capture width ratio of the WEC and the wave energy contribution given against the peak wave period

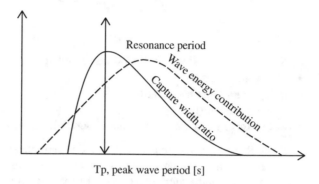

Tp, peak wave period [s]

The scaling ratio affects directly the mass and geometry of a body, which thereby influences its natural period of oscillation (more details in Chap. 6).

To give an order of magnitude of the optimal size in terms of power production for generic full-scale WECs in a location with relatively high wave power level:

- Point absorbers will be in the size of 5–15 m in diameter
- Salter's duck will have diameters in the range of 10–20 m
- Pivoting flaps will be in the range of 15–25 m wide and their thickness between 2–10 m.
- A floating overtopping WEC (or any reference structure) would optimally be at least as long as one wavelength, in order to be stable.

9.2.5.2 Scaling Law

When addressing the scaling of the mechanical interactions between fluids and solids, three main kind of forces are of importance: the inertia, gravitational and viscosity force. Depending on the case, the relative magnitude of those forces varies. Their relative importance can be quantified using two non-dimensional numbers: The Froude and Reynolds numbers [19]. Ideally, the same balance between the different forces should be maintained for the model tests as for the full-scale ones.

As inertia forces are normally predominant for the scaling of the body-fluid interaction of WECs, the Froude's Model Law (Fr) is used to transfer data between different scales. It ensures the correct similarity in between geometrical, kinematical and dynamical features. Froude's scaling law can be summarized as:

$$Fr = \sqrt{\frac{Inertia\,forces}{gravity\,forces}} = \frac{U}{\sqrt{gL}} \tag{9.18}$$

$$\frac{U_m}{\sqrt{gL_m}} = \frac{U_f}{gL_f} = \geq Uf = Um\sqrt{S} \tag{9.19}$$

where:

- $S = L_f/L_m$ = scaling ratio, requiring geometrical similarity (–)
- U = velocity (m/s)
- g = gravitational acceleration (m/s^2)
- L = dimension (m)
- Subscripts m and f stand for model and full-scale

Table 9.3 presents more explicitly the direct application of Froude's Model Law for scaling lab model related characteristics and results. The column presenting an example, presents the multiplication factor that has to be used on the model results for the different parameters to obtain the full scale, equivalents.

Table 9.3 Scaling of parameters following Froude's scaling law

Parameter	Unit	Scaling ratio	Example of scaling by 1:20
Length	m	S	$1 \to 20$
Area	m^2	S^2	$1 \to 400$
Volume	m^3	S^3	$1 \to 8000$
Time	s	S$^{0.5}$	$1 \to \sqrt{20}$
Velocity	m/s	S$^{0.5}$	$1 \to \sqrt{20}$
Force	N	S^3	$1 \to 8000$
Power	W	S$^{3.5}$	$1 \to 35777$

Note that whenever possible, test results should be expressed as non-dimensional values, meaning that they are applicable for different scaling ratios e.g. the capture width ratio of a device.

Scaling of other (non-inertia dominating) parameters depend on other specific scaling laws, meaning that for example dimensions are not scaled on the same way as compressibility or as friction. This makes it particularly difficult to scale systems such as OWC's or PTO systems (more details in Chap. 6). Thereby, it can be very difficult to scale systems accurately, as each system need to comply with the scaling laws in order to be representative.

9.2.5.3 Optimising the Scaling Ratio

Before starting the tank test campaign, a scaling ratio needs to be defined in order to scale the sea states. In practice, the offshore wave conditions and the specifications of the wave tank are fixed, while an estimation of the size of the full-scale WEC, should be made. A suitable scaling ratio needs to be found, that allows a decent representation of these scaled wave conditions in the tank, and a model needs then the be made following the same scaling ratio. Note that a larger scaling ratio corresponds to smaller scaled wave conditions, water depth and model.

An illustration of the main power performance result of tank tests is presented in Fig. 9.9. It contains the non-dimensional performance (η or CWR) for all the sea states tested with a fitted curve through the results. The results from this test campaign, can then be used as well for other scaling ratios, based on approximations with the fitted curve.

In the following figures, an example is given of the effect of having a too small or too large scaling ratio on the performance and the maximum power production (based on the values given in Table 9.2). As mentioned before, by adapting the scaling ratio to the tank test wave conditions are modified and the resulting non-dimensional performance can be obtained from the fitted power curve. In between the 3 case, the curve is translated in function of the scaling ratio; while off course the full scale wave conditions and thereby the wave energy contribution curve remain the same. Note that the wave energy contribution is usually the largest for the average wave conditions, while they decrease for the largest and smallest ones (Fig. 9.14).

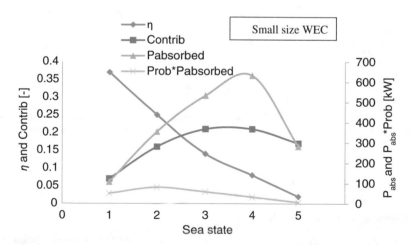

Fig. 9.14 Illustration of the possible effects of the scaling ratio, with as reference the lab model, on the performance of the device. In this figure, the curves for the smallest size of the WEC are presented

Figure 9.14 illustrates that a too small scaling ratio will lead to a peak in the non-dimensional performance curve (η, corresponding to the resonance period of the wave activated body) well below the peak of the wave energy contribution curve (*contrib*). In this case, η decreases and the wave power content increases with the sea states. Although the peak on the wave energy contribution is found for sea state 3 and 4, the maximum absorbed power contribution (Pabs x Prob) is found at sea state 2. This mismatch indicates that the resource could be better exploited. The resulting values for this scaling ratio are:

- $P_{abs, \, overall}$ of 226 kW
- $\eta_{overall}$ is 0.116
- Capacity factor is approximately 0.36.

In the second case (Fig. 9.15), the size of the device is enlarged, which results in a peak of the η curve at sea state 3, which is close to the maximum wave energy contribution. This has an immense effect on the absorbed power in the sea states, as here η increases with the wave power in the first 3 sea states, however it also comes at a large cost of the capacity factor. The results are:

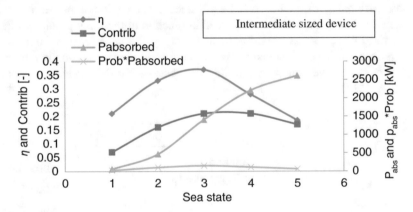

Fig. 9.15 Illustration of the possible effects of the scaling ratio, with as reference the lab model, on the performance of the device. In this figure, the curves for the intermediate size of the WEC are presented

- $P_{abs, \, overall}$ increases to 425 kW
- $\eta_{overall}$ to 0.218
- Capacity factor approximately 0.13

Although the large gain in power production, the capacity factor is significantly reduced as the maximum $P_{abs, \, ss}$ is approximately five times larger, while $P_{abs, \, overall}$ is only approximately twice larger. This indicates that although the energy production has significantly increased, the PTO system and structure got larger and thereby more expensive, which might not always result in a more cost-effective solution

In the last figure of the illustration (Fig. 9.16), the device is even further increased in size, which makes the peak of the ND performance curve to coincide with sea state 5. In this case:

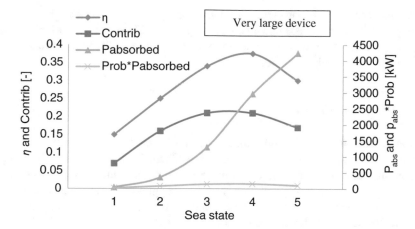

Fig. 9.16 Illustration of the possible effects of the scaling ratio, with as reference the lab model, on the performance of the device. In this figure, the curves for the largest size of the WEC are presented

- $P_{abs,\ overall}$ decreased to 366 kW
- $\eta_{overall}$ decreased to 0.188
- Capacity factor is even further diminished.

In this example, the optimal scaling ratio will most likely be slightly lower than the intermediate sized WEC, as there will be the best compromise between energy production and size of the device and its PTO system. However, a complete cost model should be included in order to find the most cost effective size of the structure including installation, power connection mooring and maintenance. This requires also that the structural design of the scaled models can handle the scaled loads of the extreme conditions (see more in Chap. 4).

9.2.6 Structural and Mooring Loads

9.2.6.1 Introduction

The objective of structural and mooring load tests is to obtain a good sense of the maximum loads that can be expected on these parts of the system. These tests need therefore to be executed on a very representative model and in the wave conditions resulting in these maximum loads (design wave conditions), which usually correspond to the extreme wave conditions. A set of extreme wave conditions, combining wave, current and wind specifications are provided in relevant design standards, such as the DNV standards on offshore wave and wind energy [10, 12, 15].

Before being able to obtain these resulting loads, usually the mooring configuration needs to be optimised. This will physically correspond to adapting the force-displacement response of the mooring system, which can be done by changing the length, mass, buoyancy and other possible parameters of its different members. Once the mooring configuration is optimised, then the structural and mooring forces can be obtained.

9.2.6.2 Mooring Forces

A rough estimation of the mooring forces can be obtained by various numerical means, e.g. with the Morison equation (see Chap. 7). For a given water depth, maximum excursion of the WEC and the maximum horizontal mooring force at the WEC, a suitable mooring solution can be designed. A static analysis of the force-displacement curve can then be calculated, against which the experimental tests will provide the dynamic analysis.

As the motions and the mooring forces are strongly influenced by the mooring stiffness, the mooring stiffness, and thereby the force-displacement curve, needs to be scaled accordingly. In practice, if a catenary mooring system is used then it can be possible to directly scale the mass/unit length of the chain and the geometry and mass of a buoy. If elastic properties of the system are of importance then the selection of the material is important and the right stiffness can be obtained (Fig. 9.17).

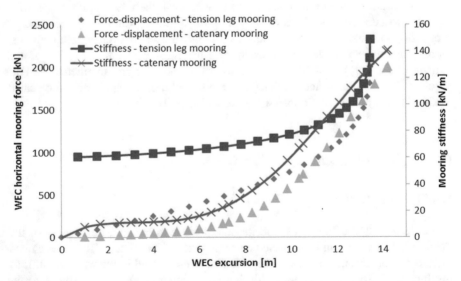

Fig. 9.17 Example of the force displacement—curves and stiffness relative to displacement curves for a catenary and tension leg mooring system [20]

Note that the surge excursion of the WEC will be strongly affected by wave groups, even more than from one large individual wave. It is therefore important to repeat the same design wave conditions with different wave trains.

9.2.6.3 Structural Design and PTO Loads

Each component of the WEC requires to be dimensioned in function of the highest loads (e.g. 1/250 force) and adequate safety factors. Multiple sensors need thereby to be incorporated at strategic places in the structure and other components, so that decent measurements of the maximum loads can be made (Fig. 9.18).

Fig. 9.18 Picture of a floater of the Wavestar WEC being experimentally tested for wave loads. Courtesy of Wavestar

It is highly recommended to complement these experimental tests with additional calculations or numerical models.

9.2.7 Parametric Study

9.2.7.1 Physical Alterations to the Model

The objective of making changes to the model is often to investigate ways to increase its performance, or to test other hydrodynamic, structural or more economically design solutions.

It can be sufficient to perform only a few trials to assess the impact of alterations. If the alteration is not tested in all operational or design sea states, it should especially be investigated in the conditions where it is the most critical, e.g. where the wave energy contribution is the highest or where the model obtains its highest loads. The results of the test should be compared to a reference tests, being the original setup without any alterations. Every alteration should be investigated separately in order to identify the influence of every parameter individually.

9.2.7.2 Modification of Wave Parameters

In a laboratory environment, normally long crested irregular waves are used that are defined by an H_s and T_p values together with a defined wave spectrum, which corresponds to a sea state. However, the environmental parameters describing a marine location (wind, currents, bathymetry, directional wave spectrum, …) are numerous and can in many cases vary independently from each other. Therefore, it is of importance to asses to influence of each of these relevant environmental parameters in operational and design wave conditions.

The influence of these additional environmental parameters needs to be addressed separately and over the whole range of their possible extent, which can then be compared with the original reference situation.

9.3 Sea Trials

9.3.1 Introduction

After extensive tank testing and individual components analysis on test benches, the first sea trials marks the beginning of a new very exciting but demanding stage in the development of a WEC. It is initiated by an intense preparation effort, requiring to investigate a vast range of new grounds and challenges, just to make everything ready and to be prepared for the new uncontrolled environment with restricted access. Sea trials can be performed for a wide range of objectives, which can have a strong influence on the capabilities of the WEC prototype and on the test location. Besides the data gained from the sea trials, the construction of WEC and the experience with its operation and maintenance, are as well highly valuable. Some of the main objectives for sea trials can be:

- To demonstrate the technology in real ocean conditions.
- To (ultimately) operate the system as an autonomous power plant.
- To measure, verify and validate loads, motions and power performance calculations and estimations.
- To refine the Levelized Cost of Energy (LCoE) estimations, based on the new and more representative mean annual energy production (MAEP), capital costs (CapEx) and operation and maintenance costs (OpEx) evaluations.

Based on the sea trials, many cost and power performance estimations will be made for a commercially operated WEC array. However, there will be some significant differences between the situation of the sea trial and of the WEC array, in terms of design and size of the WEC, environmental conditions and array effects. An overview of the situation is presented in Table 9.4.

Table 9.4 Estimating of costs and power production of a commercially operated WEC, based on sea trials

	Source: sea trial		Estimation: WEC array
# of WECs	1 or several		Multiple
Scale	≈1:6–1:1	→	Full-scale
Location	Test site (site 1)		WEC array site (site 2)

The influences between the two situation can have significant influence on the mean annual energy production and cost of the WEC. Therefore, they need to be carefully investigated, possibly during the sea trials, but otherwise complemented by representative model tests or validated numerical models.

In this chapter, especially the power performance evaluation is emphasized. The presented methodology can be applied to all WECs and its aim is to estimate the electricity production of a full-scale WEC (array), operating as a power plant, at another location, based on the measurements of the sea trials. Methods for this are being currently drafted under the IEC 62600-102/CD [21].

9.3.2 Performance Assessment of WECs Based on Sea Trials

9.3.2.1 Introduction

A condensed overview of a methodology to equitably assess the performance of wave energy converters based on sea trials will be presented here [22, 23] and case studies of it can be found for the Pico OWC and the Wave Dragon WEC [24, 25].

The "Equitable Performance Assessment and Presentation" methodology aims at assessing the performance of any device, based on sea trials, in a transparent and equitable way, resulting in an estimation of the mean annual energy production (*MAEP*) together with a corresponding accuracy.

Sea trials are (generally) very expensive and time-consuming, as they require heavy equipment and some wave conditions only occur sporadically [1]. Moreover, various problems might occur and different parameters have to be tested and optimized. This (usually) leads to a vast amount of discrepancy in the recorded performance of the device, which each should be clearly marked—especially in the early stages of testing. The methodology thereby accepts incomplete and interrupted data series from sea trials, which were not obtained during autonomous mode. It, however, expects the developer to provide clear and transparent information regarding the data and the sea trials. On the contrary, for (near to) commercial devices all the data recorded over a given time period under continuous and autonomous operation of the WEC would need to be used, without exceptions (IEC 62600-100/TS) [26].

A robust but flexible methodology is required that can take the discrepancy of the power performance into account, while enabling the estimation of the MAEP of the device at the test location and of the full-scale device at any location of interest. The methodology favours larger data sets as it makes the resulting performance more robust and the corresponding uncertainty interval smaller.

An overview of the methodology is given in Fig. 9.19.

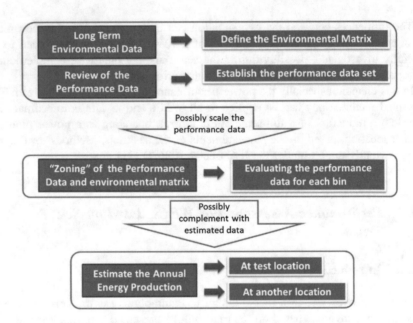

Fig. 9.19 Overview of the power performance assessment procedure

9.3.2.2 Preparing the Environmental and Performance Data

The first part is to process separately the power performance data and the environmental data of the test and possibly of another given location.

In order for the environmental data to be representative, it requires to cover a long period of time (>10 year) and it can be measured or hindcasted. This will usually be condensed into a bi-variate scatter diagram ($H_s - T_e$) in order to represent the wave conditions. However, this can be extended to more detailed (n + 2) scatter diagrams by including other environmental parameters, such as e.g. wave direction, as they can be of significant influence on the power performance of the WEC. In case that the MAEP is calculated for another location, it will be needed to take the environmental parameters that are different and that have an influence on the performance of the WEC into account in the description of the environmental conditions. This could also be important, if the environmental conditions during the sea trials have not been representative for the long-term average conditions (Fig. 9.20).

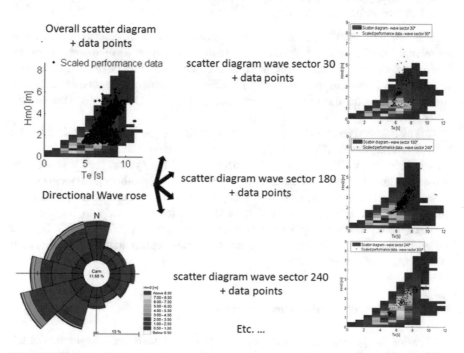

Fig. 9.20 The overall bi-variate scatter diagram with data points and corresponding directional wave rose (*left*) or the directional scatter diagrams with corresponding data points

The same method can be used if the environmental matrix is represented more in detail than just by a bi-variate scatter diagram. The performance data will then have to be divided over the different scatter diagrams and the probability of occurrence of these different scatter diagrams will have to be known. Afterwards, their results can be added to each other to obtain the overall MAEP.

Besides the long-term environmental data, there will also be the performance and environmental data that is collected during the sea trials. This recorded data, referred to as the "performance data", needs to include a wide range of environmental and device dependent parameters that are evaluated over a defined timespan, usually 30 min [27]. The list of parameters to be included depends on the desired application and is especially large whenever a parametric study or a wave-to-wire analysis is intended to be produced. The key and indispensable environmental and performance parameters required for each data sample are:

- H_{m0} Significant wave height derived from spectral moments
- T_e Wave energy period
- P_w Wave power (taking the water depth into account)
- P_{el} Produced electrical power
- P_{abs} Primary absorbed power by the device
- η Capture width ratio (*CWR*)

This list will usually be broaden by various parameters describing the environmental conditions more specifically, such as e.g. the wave direction, or by parameters describing the configuration and setup of the device, e.g. control strategy. It is important to include any possible environmental parameter that has a strong influence on the power performance of the WEC (i.e. wind, current and water level).

It is also desirable to have a measurement of the power directly absorbed by the WEC from the waves, P_{abs}, without further power conversion modules in between. This is because the P_{abs} represent the upper limit of the system (and it is scalable and can be used to define the efficiency of the PTO). These other components in the PTO could possibly also be changed or improved afterwards and could possibly be difficult to scale. However, for full-scale devices that are ready to be commercialized, the representative power performance measurement will most-likely be at the grid connection, as it will give the most accurate representation and estimation of the MAEP of the whole device.

9.3.2.3 Scaling of the Performance Data

In case that the power performance analysis has to be made for a larger size of the WEC, than tested during the sea trials, then the environmental parameters of the performance data can possibly be scaled at this moment of the procedure. Froude scaling should be used, as stipulated in Sect. 9.2.5, on the environmental parameters, while only the relative measures for the power performance parameters (capture width ratio) can then kept being used.

The optimal size of the WEC can depend on various parameters, from hydrodynamics to economics and logistics, on which a brief approach on the hydrodynamic optimization is given in Sect. 9.2.5.3.

9.3.2.4 Categorising the Data

All the performance data collected during an one straight operational period of the WEC, without interruption or incomplete data, will be used to asses the overall performance. The more data that is available, the more robust the estimation will be and thereby the lower the uncertainty related to the obtained performance of the WEC in real sea conditions. The sea trial experience could as well give an indication on the expected availability of the WEC, which as well has an influence on the MAEP.

The relevant performance data will be categorised into subsets, according to the definition of the bins of the scatter diagram and possibly also according to the abundance of performance data. Note that in some cases, only a scatter diagram will be available for a given location (and not the long-term timeseries); here the same bins will have to be used as the one defined in the scatter diagram.

An illustration of these bins dividing the data into subsets is presented in Fig. 9.21. For each of these bins, corresponding to data subsets, an average *CWR* and related uncertainty value will calculated.

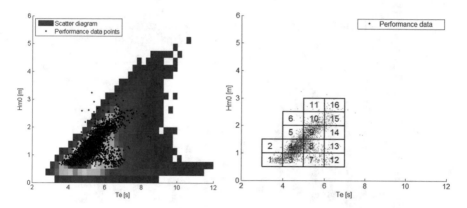

Fig. 9.21 Scatter diagram with performance data (*left*) and zoning of the performance data (*right*)

The bins are delimited by a certain range of wave height and wave period (H_{m0} and T_e), which is suggested to be the same for all of them. This corresponds normally to bins of 0.5 or 1 m in significant wave height and 0.5, 1 or 1.5 s in wave period. In practice, the size of the bins, will influence the resolution of the power matrix, the amount of data points that will be found in a bin, the variation between the performance of different data points and thereby the uncertainty related to the average power performance of a bin.

So far, no standard selection criterion exist (or been completed) as not enough developers have shared their results and approaches, however some specifications can be suggested:

- The bins should all have the same size.
- The performance data analysis has to be done with the respective η value and with the absolute performance in *kW*.
- At least five performance data points have to be included in the data selection of each bin. However, it is strongly encouraged to increase the amount significantly when sufficient data is available, in order to obtain a more robust performance representation.
- All the acquired performance data in a certain period should be used to represent the overall performance. This can be difficult for new prototypes undergoing sea

trials, which did not acquire sufficient data during long enough autonomous operation of the WEC. Exceptions to this rule could be then accepted (meaning that the developer choses to only include a subset of the available performance data, by screening the performance data), as long as this is clearly stated and done on a transparent manner. This will indicate that the WEC is still a prototype version (Fig. 9.22).

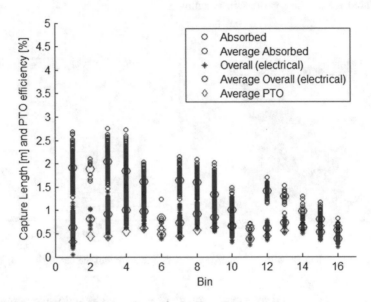

Fig. 9.22 Representation of the performance data for all bins

For each of the bins representative wave and performance have to be calculated. The parameters characterizing the wave conditions (H_{m0}, T_e), probability of occurrence ($Prob_{bin}$) and the wave energy contribution ($Contrib_{bin}$) have to be derived from the long term environmental data, given by Eqs. (9.3) and (9.4).

The average and uncertainty of the performance of the WEC for a bin has to be based on the performance data. This is given in terms of a non-dimensional performance value (η_{bin}) and the uncertainty is expressed in terms of sample standard deviation (s_{bin}) and confidence interval (CI_{bin}), using a standard confidence level at 95 % and a Student's t-distribution. The distribution might not be the most suitable and can be adapted in order to be more representative and accurate.

This approach incites the WEC operation to focus on demonstrating good performance over longer periods of time (resulting in a greater amount of performance data points) in order to stabilise the η_{bin} and to reduce the CI_{bin}. The average η and its corresponding confidence interval for each bin, based on the selected performance data points (n), can be calculated as such:

$$\eta_{bin} \mp CI_{bin} = \frac{\sum_{i=1}^{n} \eta_i}{n} \pm t^* \frac{s}{\sqrt{n}} \tag{9.20}$$

9.3.2.5 Complementing the Performance Data

As mentioned before, at the end of the sea trials there might not be sufficient performance data to cover the whole scatter diagram abundantly, as the sea trial period is limited, some wave conditions only occur infrequently, and the WEC might not always be in operation. Therefore, it is likely that some bins of the scatter diagram might not be populate with sufficient performance data in order to calculate for it a representative performance value.

The power performance of the WEC might be estimated based on the measured performance data from the sea trials through validated numerical models or experience from tank testing. This off course has to be done very carefully, and therefor these estimations have to be very conservative (Fig. 9.23).

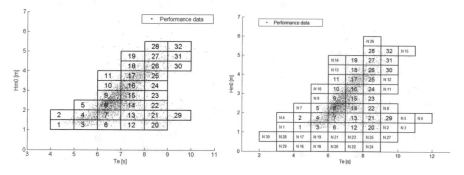

Fig. 9.23 The original (*left*) and extended (*right*) bins zoning the power performance data (*little dots*) obtained through sea trials and estimations

Note that the use of estimated power performance data, has to be explained and clearly stated together with the mean annual energy production estimation.

9.3.2.6 Estimating the MAEP

The P_{abs} by the WEC for the environmental conditions linked to each bin can be obtained by multiplied their non-dimensional performance values by their corresponding P_w. This will in other words lead to the absorbed power matrix.

The absorbed power matrix can then be multiplied by the PTO efficiency matrix, which can be derived from the sea trials, in order to obtain the electrical power matrix. Note that the PTO efficiency matrix could be updated, to take improvements of the system into account or differences due to scaling. This is acceptable as long as this is explained, clearly indicated and that the changes are conservative (Fig. 9.24).

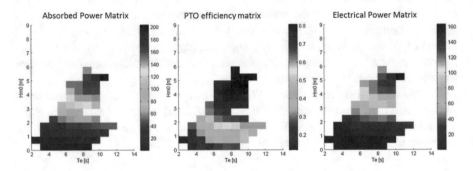

Fig. 9.24 The absorbed power matrix, the PTO efficiency matrix and the resulting electrical power performance matrix

The resulting mean annual energy production, with its uncertainty, can be calculated similarly as in the case of the performance estimation based on experimental tank testing (Sect. 9.2.4.2). It corresponds to making the sum of the product of the electrical power matrix with the scatter diagram [8, 28].

The unbiased estimate of the overall standard deviation ($s_{overall}$) and confidence interval ($CI_{overall}$) can be obtained with Eq. (9.21), in which X can be replaced by s or CI in order to obtain their individual overall appreciation:

$$X_{overall} = \sqrt{\sum_{bin=1}^{n} \left(\eta_{bin}^2 + X_{bin}^2\right) \cdot Contrib_{bin} - \left(\sum_{bin=1}^{n} \eta_{bin} \cdot Contrib_{bin}\right)^2} \quad (9.21)$$

The estimation of the annual energy production together with an estimation of its uncertainty will give a strong indication of the accuracy and technological readiness of the WEC.

References

1. Holmes, B.: OCEAN ENERGY : Development and Evaluation Protocol—Part 1 : Wave Power. HMRC—Marine Institute, pp. 1–25 (2003)
2. Meyer, N.I., Rambøll: Bølgekraftprogram - Afsluttende rapport fra Energistyrelsens Rådgivende Bølgekraftudvalg (2002)

3. Kofoed, J.P., Frigaard, P.: Development of wave energy devices: the Danish case. J. Ocean Technol. **4**(2), 83–96 (2009)
4. Salcedo, F., Rodriguez, R., Ricci, P., Santos, M.: OCEANTEC : sea trials of a quarter scale prototype. In: 8th European Wave and Tidal Energy Conference (EWTEC), Uppsala, Sweden (2009)
5. O'Sullivan, D.L., Lewis, A.W.: Generator selection and comparative performance in offshore oscillating water column ocean wave energy converters. IEEE Trans. Energy Convers. **26**(2), 603–614 (2011)
6. Salter, S.H., Taylor, J.R.M.: Edinburgh wave power project—bending moments in long spines. United Kingdom Department of Energy, pp. 1–233 (1984)
7. Nielsen, K.: Development of recommended practices for testing and evaluating ocean energy systems. OES IA Annex 2—Doc. No T02—0.0 (2010)
8. Pecher, A.: Performance evaluation of wave energy converters. Aalborg University (2012)
9. Kofoed, J.P., Folley, M.: (2016) Determining mean annual energy production. in Folley M. (ed.) Numerical Modelling of Wave Energy Converters: State-of-the-Art Techniques for Single Devices and Arrays, pp. 253-266, 1st edn. Academic Press, Incorporated
10. Det Norske Veritas: DNV-RP-C205: environmental conditions and environmental loads (2010)
11. Liu, Z., Frigaard, P.: Generation and analysis of random waves (1999)
12. Det Norske Veritas: DNV-OS-J101: design of offshore wind turbine structures (2013)
13. Brodtkorb, P.A., Johannesson, P., Lindgren, G., Rychlik, I., Rydaan, J., Sjap, E.: WAFO: a Matlab toolbox for analysis of random waves and loads. In: Proceedings of the 10th International Offshore and Polar Engineering Conference, Seattle, USA, pp. 343–350 (2000)
14. DHI: Matlab toolbox (2012)
15. Det Norske Veritas: DNV-OS-E301: position mooring (2010)
16. Holmes, B.: Tank testing of wave energy conversion systems. The European Marine Energy Centre Ltd (Marine Renewable Energy Guides), pp. 1–82 (2009)
17. Kofoed, J.P., Antonishen, M.: The crest wing wave energy device—2nd phase testing. Aalborg University DCE Contract Report No. 59, pp. 1–37 (2009)
18. Pecher, A., Kofoed, J.P., Larsen, T.: Design specifications for the Hanstholm WEPTOS wave energy converter. Energies **5**(4), 1001–1017 (2012)
19. Payne, G.: Guidance for the experimental tank testing of wave energy converters. Supergen Mar., pp. 1–47 (2008)
20. Pecher, A., Foglia, A., Kofoed, J.: Comparison and sensitivity investigations of a CALM and SALM type mooring system for wave energy converters. J. Mar. Sci. Eng. **2**(1), 93–122 (2014)
21. IEC TC114/TS62600-102: Marine energy—wave, tidal and other water current converters—part 102 : power performance assessment of electricity producing wave energy converters. http://www.iec.ch/dyn/www/f?p=103:23:0::::FSP_ORG_ID:1316
22. Kofoed, J.P., Pecher, A., Margheritini, L., Antonishen, M., Bittencourt-ferreira, C., Holmes, B., Retzler, C., Berthelsen, K., Le Crom, I., Neumann, F., Johnstone, C., McCombes, T., Meyers, L.E.: A methodology for equitable performance assessment and presentation of wave energy converters based on sea trials. Renew. Energy (2013)
23. Kofoed, J.P., Pecher, A., Margheritini, L., Holmes, B., McCombes, T., Johnstone, C.M., Bittencourt-ferreira, C., Retzler, C., Myers, L.E.: Data analysis and presentation to quantify uncertainty. EquiMar Protocols—Equitable Testing and Evaluation of Marine Energy Extraction Devices in terms of Performance, Cost and Environmental Impact (2011)
24. Pecher, A., Crom, I. Le, Kofoed, J.P., Neumann, F., Azevedo, E.D.B.: Performance assessment of the Pico OWC power plant following the EquiMar methodology. In: 21th International Offshore (Ocean) and Polar Engineering Conference (ISOPE), Maui, Hawaii (2011)
25. Parmeggiani, S., Chozas, J. F., Pecher, A., Friis-Madsen, E., Sørensen, H.C., Kofoed, J.P.: Performance assessment of the wave dragon wave energy converter based on the EquiMar methodology. In: 9th European Wave and Tidal Energy Conference (EWTEC) (2011)

26. IEC TC114/TS62600-100: Marine energy—wave, tidal and other water current converters—part 100 : power performance assessment of electricity producing wave energy converters. IEC (2012)
27. McCombes, T., Johnstone, C.M., Holmes, B., Myers, L.E., Bahaj, A.S., Kofoed, J.P.: Best practice for tank testing of small marine energy devices. EquiMar Protocols: Equitable Testing and Evaluation of Marine Energy Extraction Devices in terms of Performance, Cost and Environmental Impact (2011)
28. Kofoed, J.P., Pecher, A., Margheritini, L., Antonishen, M., Bittencourt, C., Holmes, B., Retzler, C.: A methodology for equitable performance assessment and presentation of wave energy converters based on sea trials. Renew. Energy **52**(2013), 990-110 (2013)

Chapter 10
Wave-to-Wire Modelling of WECs

Marco Alves

10.1 Introduction

Numerical modelling of wave energy converters (WECs) of the wave activated body type (WAB, see Chap. 2) is based on *Newton*'s second law, which states that the inertial force is balanced by all forces acting on the WEC's captor. These forces are usually split into hydrodynamic and external loads.

In general, the **hydrodynamic source** comprises the (more details in Chap. 6):

- **Hydrostatic force** caused by the variation of the captor submergence due to its oscillatory motion under a hydrostatic pressure distribution,
- **Excitation loads** due to the action of the incident waves on a motionless captor,
- **Radiation force** corresponding to the force experienced by the captor due to the change in the pressure field as result of the fluid displaced by its own oscillatory movement, in the absence of an incident wave field.

Depending on the type of WEC, the **external source** may include the loads induced by the

- **Power-take-off** (PTO) equipment, which converts mechanical energy (captor motions) into electricity (more details in Chap. 8),
- **Mooring system,** responsible for the WEC station-keeping (more details in Chap. 7),
- **End-stop** mechanism, used to decelerate the captor at the end of its stroke in order to dissipate the kinetic energy gently, and therefore avoid mechanical damage to the device.

M. Alves (✉)
Rua Dom Jerónimo Osório 11, 1400-119 Lisbon, Portugal
e-mail: marco@wavec.org

© The Author(s) 2017
A. Pecher and J.P. Kofoed (eds.), *Handbook of Ocean Wave Energy*,
Ocean Engineering & Oceanography 7, DOI 10.1007/978-3-319-39889-1_10

The hydrodynamic modelling of the interaction between ocean waves (see Chap. 3) and WECs is often split into three different phases according to the sea conditions:

(i) During small to moderate sea states linear wave approximations are valid, corresponding to the current state-of-the-art methods of hydrodynamic modelling.

(ii) Under moderate to extreme waves, in general, some sort of non-linear hydrodynamic modelling is required in order to more accurately model the wave/device interaction.

(iii) Ultimately, under stormy conditions, a fully non-linear approach is necessary to model the hydrodynamic interaction of the waves and the device.

With respect to the modelling of the external loads it is commonly accepted that the production of energy should be restricted to non-stormy conditions (WEC operating mode), comprising both small to moderate and moderate to extreme waves, which correspond to low and intermediate energetic sea states. Under stormy conditions, usually it is not necessary to model the dynamics of the PTO equipment as the WEC is interacting with extreme waves and so it must assume the survivability mode with no energy production. In the operating mode of the device the loads induced by the **PTO equipment,** the **mooring system** and the **end-stop mechanism** may be linearized under certain assumptions; however, typically they exhibit strongly nonlinear behaviour, which requires a time domain approach in order to be described properly.

Although the scope of this chapter is confined to wave-to-wire modelling it is important to emphasize that there are other modelling methods and that the most adequate one depends on several factors such as the required accuracy (which is typically inversely proportional to the computational time), the sea state (stormy or non-stormy conditions), the device regime (operational or survival mode) and its work principle (some concepts exhibit more non-linear behaviours). In view of that, the modelling tools are typically split into 3 different types:

1 **Frequency models**: The hydrodynamic interaction between WECs and ocean waves is a complex high-order non-linear process, which, under some particular conditions, might be simplified. This is the case for waves and device oscillatory motions of small-amplitude. In this case the hydrodynamic problem is well characterised by a linear approach. Therefore, in such a framework (which is normally fairly acceptable throughout the device's operational regime), and with linear forces imposed by both the PTO and the anchoring system, the first step to model the WEC dynamics is traditionally carried out in the frequency domain (where the excitation is of a simple harmonic form). Consequently, all the physical quantities vary sinusoidally with time, according to the frequency of the incident wave. Under these circumstances, the equations of motion become a linear system that may be solved in a straightforward manner.

Although frequency models have limited applicability, being restricted to linear problems where the superposition principle is valid, the frequency domain approach is extremely useful as it allows for a relatively simple and fast assessment of the WEC performance, under the aforementioned conditions. Hence, this approach is generally used to optimise the geometry of WECs in order to maximize the energy capture [1–3].

2   **Wave-to-wire models** (time domain tools): Besides the interest of the frequency domain approach, in many practical cases the WEC dynamics has some parts that are strongly non-linear, and so the superposition principle is no longer applicable. These nonlinearities arise mostly from the dynamics of the mooring system, the PTO equipment and control strategy and, when present, the end-stop mechanism. Furthermore, under moderate to extreme waves, nonlinear effects in the wave/device hydrodynamic interaction are more relevant. This requires some sort of non-linear modelling that typically consists of treating the buoyancy and the excitation loads as non-linear terms. In addition, second-order slow drift forces may be also included in a time domain description of the WEC dynamics (this force must be undertaken by the station-keeping system). To properly account for these nonlinearities the WEC modelling has to be performed in time domain. Moreover, the motion of the free surface in a sea state rarely reaches steady-state conditions, and so must also be represented in the time domain.

The time domain approach is a reasonably detailed and accurate description of the WEC dynamics. Since this approach allows modelling of the entire chain of energy conversion from the wave/device hydrodynamic interaction to feeding into the electrical grid, time domain models are commonly named wave-to-wire codes. The most relevant outcomes of a wave-to-wire code includes, among others, estimates of the instantaneous power produced under irregular sea states, motions/velocities/accelerations of the WEC captor and loads on the WEC. Besides, wave-to-wire models are extremely useful tools to optimize the WEC control strategy in order to maximize the power captured. The Structural Design of Wave Energy Devices (SDWED) project, led by Aalborg University, has generated a comprehensive set of free software tools including advanced hydrodynamic models, spectral fatigue models and wave to wire models [4].

3   **Computational fluid dynamics—CFD**: Due to the large computational time the use of CFD codes is typically restricted to study the wave/device interaction under extreme waves, which is a strongly non-linear phenomena. Normally, the main objective in this case is to model the WEC dynamics in its survival mode with no energy production (in order to evaluate the suitability of the survival strategy). This type of wave-body interaction is usually computed solving the Reynolds Averaged Navier-Stokes Equations (RANSE[1]) with some

---

[1]The decomposing of the Navier-Stokes equations into the Reynolds-averaged Navier–Stokes equations (RANSE) makes it possible to model complex flows, such as the flow around a wave power device. RANSE are based on the assumption that the time-dependent turbulent velocity fluctuations may be separated from the mean flow velocity. This assumption introduces a set of unknowns, named the Reynolds stresses (functions of the velocity fluctuations), which require a turbulence model to produce a closed system of solvable equations.

sort of numerical technique to model the free surface of the water. Among several different methods to model the free surface one of the most commonly used is the Volume of Fluid (VoF) [5]. At present there are some CFD codes capable of modelling this sort of wave-body interaction and flows with complex free-surface phenomena such as wave breaking and overtopping (see Sect. 10.3: Benchmark Analysis).

10.2 Wave-to-Wire Models

At present there are many designs being pursued by developers to harness wave power, which may be categorized according to the location and depth in which they are designed to operate, i.e. shoreline, near shore or offshore, or by the type of power capture mechanism. However, there is no common device categorization that has been widely accepted within the international research and technology development community, but the most popular distinguishing criteria is based on their operational principle. According to this criterion WECs are usually divided into six distinct classes: attenuators; point absorbers; oscillating-wave surge converters; oscillating water columns (OWC); overtopping devices; and submerged pressure-differential devices [6]. These categories may be regrouped into three fundamentally different classes, namely OWC, WECs with wave-induced relative motions and overtopping devices. For WECs within the two first fundamental classes the generic approach to develop wave-to-wire models presented herein is valid, however, for overtopping concepts the performance analysis requires the use different type of numerical tools based on empirical expressions (such as e.g. WOPSim: Wave Overtopping Power Simulation [7]) or CFD codes.

In the field of wave energy, the term wave-to-wire refers to numerical tools that are able to model the entire chain of energy conversion from the hydrodynamic interaction between the ocean waves and the WEC to the electricity feed into the grid. In terms of complexity, and consequently time expenditure, these types of numerical tools are in-between frequency domain codes, which are much faster but less accurate (because all the forces are linearized), and CFD codes, which are currently the most precise numerical tools available, but also extremely time demanding, which makes their use unviable to solve the majority of problems in this field.

This section presents a discussion on the assumptions, considerations and techniques commonly used in developing wave-to-wire models, highlighting the limitations and the range of validity of this type of modelling tool. A general discussion is presented aiming to embrace the majority of existing WECs, nevertheless when appropriate, an annotation regarding the fundamental differences in the working principle of some particular WECs and the subsequent adjustments in the wave-to-wire model will be made.

10.2.1 Equation of Motion

In essence, the algorithm to build a wave-to-wire model relies on Newton's second law of motion, which states that the inertial force is balanced by all of the forces acting on the WEC's captor. This statement is expressed by the equation

$$M\ddot{\xi}(t) = F_e(t) + F_r(t) + F_{hs}(t) + F_f(t) + F_{pto}(t) + F_m(t), \tag{10.1}$$

where M represents the mass matrix and $\ddot{\xi}$ the acceleration vector of the WEC. The terms on the right hand side of Eq. 10.1 correspond to:

- The excitation loads—F_e
- The hydrostatic force—F_{hs}
- The friction force—F_f
- The radiation force—F_r
- The PTO loads—F_{pto}
- The mooring loads—F_m

In the following section a discussion on the different sources of loads on the WEC captor is presented and their impact on the overall dynamics of the WEC is given in order to substantiate the assumptions and simplifications commonly considered in the development of wave-to-wire codes.

10.2.2 Excitation Force

The **excitation force** results from the pressure exerted on the body's wetted surface due to the action of the incoming waves. The most popular approach to compute this force is based on linear wave theory, in which the body is assumed to be stationary and the area of the wetted surface constant and equal to the value in undisturbed conditions. Obviously this assumption is only valid for small wave amplitudes, which is a fundamental assumption of linear theory. Therefore, under linear assumptions the excitation load on the WEC captor is given by

$$F_{exc}(t) = \int_{-\infty}^{\infty} f_{exc}(t - \tau)\eta(\tau)d\tau, \tag{10.2}$$

where η is the free surface elevation due to the incident wave (undisturbed by the WEC) at the reference point where the WEC is located and f_{exc} is the so called excitation impulse response function derived from the frequency coefficients commonly obtained with a 3D radiation/diffraction code (see Sect. 10.3). Equation 10.2 shows that it is necessary to model the random sea state behaviour in order to estimate the excitation force. The most common approach consists of using Airy wave theory, a linear theory for the propagation of waves on the surface of a

potential flow and above a horizontal bottom. The free surface elevation, η, may be
then reproduced for a wave record with duration T as the sum of a large (theoretically infinite) number, N, of harmonic wave components (a Fourier series), the
so called wave superposition method, as

$$\eta(t) = \sum_{i=1}^{N} a_i \cos(2\pi f_i t + \alpha_i), \tag{10.3}$$

where, t is the time, a_i and α_i the amplitudes and phases of each frequency,
respectively, and $f_i = i/T$. The phases are randomly distributed between 0 and 2π,
so the phase spectrum may be disregarded. Hence, to characterize the free surface
elevation only the amplitudes of the sinusoidal components need to be identified,
which are given by

$$a_i = \sqrt{2S_f(f_i)\Delta f}, \tag{10.4}$$

where S_f is the variance density spectrum or simply energy spectrum (see Fig. 10.1)
and Δf the frequency interval. As only the frequencies f_i are presented in the energy
spectrum, while in reality all frequencies are present at sea, it is convenient to let the
frequency interval $\Delta f \to 0$. The spectrum of energy is usually plotted as energy
density, (unit of energy/unit frequency interval, Hz) given by the amount of energy
in a particular frequency interval.

Fig. 10.1 Typical variance
density spectrum

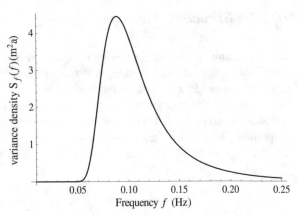

For more realistic descriptions of the wave surface elevation the wave's directionality must be considered. In this case the direction resolved spectrum $S(\beta,f)$,
dependent on the frequency, f, and wave direction, β, is written as

$$S_f(f, \beta) = D(f, \beta)S_f(f), \tag{10.5}$$

where the directional distribution $D(\beta,f)$ is normalized, satisfying the condition

$$\int_{-\pi}^{\pi} D(\beta,f)d\beta = 1. \tag{10.6}$$

The spectrum is defined with several parameters in which the most important ones are the significant wave height, denoted by H_s or $H_{1/3}$ (which corresponds to the average of the highest third of the waves), and the peak period, T_p corresponding to the period with the highest peak of the energy density spectrum (the spectrum may have more than one peak).

10.2.3 Hydrostatic Force

When a body is partially or completely immersed in a liquid it will experience an upward force (buoyancy) equal to the weight of the liquid displaced, which is known as Archimedes' principle. The **hydrostatic force** results from the difference between this upward force and the weight of the body. Accordingly, the variation of the captor submergence due to its oscillatory motion under a hydrostatic pressure distribution causes a change in the buoyancy (equal to the change of weight of displaced fluid) and hence a variation in the hydrostatic force.

A fundamental assumption of linear theory is that the resulting body motions are of small amplitude, which normally conforms with the behaviour of WECs during the operational regime. In fact, the motion of WECs tends to be of small amplitude because otherwise the dissipative viscous effects would be dominant in the device dynamics, which would ultimately limit the motion and reduce the device efficiency. Therefore, the hydrostatic force, F_{hs}, is commonly implemented in wave-to-wire models merely as a function proportional to the body displacement, where the proportionality coefficient is known as the hydrostatic coefficient, i.e.,

$$F_{hs}(t) = Gz(t), \tag{10.7}$$

where G is the hydrostatic coefficient and z the motion in the direction of the degree of freedom (DoF) being considered. In the case of several DoFs being analysed G and z represent the hydrostatic matrix and the displacement vector, respectively.

For example, in the case of a heaving body undergoing small-amplitude oscillations the variation of the buoyancy force may be simply given by

$$F_{hs}(t) = \rho g A z(t), \tag{10.8}$$

where ρ denotes the water density, g the gravitational acceleration, A the cross sectional area of the body in undisturbed conditions and z its vertical displacement. The variation of the volume of water displaced by the oscillating body is equal to the variation of its submerged volume, given by Az. We should note that typically the assumption of constant cross sectional area along the vertical axis is only valid for motions of small-amplitude. Depending on the body geometry, typically this simplification (based on the linear wave theory) is not valid for large-amplitude motions where in general the variation of the cross-sectional area is more noticeable, and so a non-linear approach is required to accurately assess the hydrostatic force.

10.2.4 Mooring Loads

Wave drift forces,[2] along with currents and wind, have a tendency to push the WEC away from the deployment position. To prevent this drifting, the WEC should be maintained in position by a station-keeping system, also commonly called a "mooring system". The station-keeping system is usually designed to withstand survival conditions, e.g. 100 year storm conditions. The moorings designed for floating WECs are required to limit their excursions and, depending on the concept, aligning its position according to the angle of incidence of the incoming waves. Moreover, unlike typical offshore structures, the mooring design has an additional requirement of ensuring efficient energy conversion, since it may change the response of the WEC and so change its ability to capture wave energy.

Depending on the working principle of the device and ultimately the manner in which the mooring system provides the restoring force, mooring systems might be passive, active or reactive. Passive mooring systems are designed for the unique purpose of station-keeping. Conversely, active mooring systems have a stronger impact on the dynamic response of WECs since the system stiffness may be used to alter the resonant properties of WECs. Ultimately, reactive mooring systems are applied when the PTO exploits the relative movements between the body and the fixed ground, such that the mooring system provides the reaction force. In this mooring configuration the inboard end of the mooring line/s is connected to the PTO equipment which controls the tensions or loosens of the mooring line/s in order to adjust the WEC position according to the established control strategy. A review of design options for mooring systems for wave energy converters is presented in Refs. [8, 9].

Mooring systems are traditionally composed of several mooring lines (slack or taut), with one extremity attached to the device, at a point called the fairlead, and the other extremity attached to a point that must be able to handle the loads applied by the device through the line. This point can be fixed to an anchor on the seabed, or moving, e.g. the fairlead on another floating offshore structure. Mooring lines are

---

[2]Drift forces are second-order low frequency wave force components. Under the influence of these forces, a floating body will carry out a steady slow drift motion in the general direction of wave propagation if it is not restrained. See further in Chap. 7.

usually composed of various sections of different materials (chain, wired-ropes, polyester, etc.). Some additional elements, such as floats or clump weights can be attached to the line to give it a special shape.

Depending on the objectives of the simulation, the mooring system can be modelled with different levels of accuracy, and thus different computational efforts. Hence, it is important to understand the level of detail and accuracy required in order to select the most appropriate modelling approach. Essentially mooring models may be split into two main categories: quasi-static and dynamic models.

Quasi-static models depend only on the position of the fairlead and the anchor at specific time-step. Therefore, they do not solve differential equations for the motion of the lines, which considerably reduces the required computational effort. Quasi-static models may be split into two types:

- **Linearized mooring model**. The most common quasi-static model is the so-called "linearized mooring model" which consists of modelling the mooring loads in the different directions of motion by a simple spring effect. The computation of the restoring effect is straightforward, but it is only effective when the device has small motions, around its undisturbed position. Whenever this approach is built-in in the wave-to-wire model it is necessary to input the mooring spring stiffness matrix (which is multiplied by the displacement vector at each time-step to define the tension at the fairlead connection).
- **Quasi-static catenary model**. The quasi-static catenary modelling approach consists of computing the tension applied by a catenary mooring line on a device using only the position of the fairlead and the anchor. This mooring modelling approach requires the inclusion of the nonlinear quasi-static catenary line equations in the wave-to-wire model, which are solved at each time-step in order to determine the value of the tension at the fairleads. This modelling approach is very simple and requires little computational effort, but it is only valid for relatively small motions about the mean position.

Quasi-static models are usually reliable to estimate the horizontal restoring effect on a device that experiences small motion amplitudes, but they are not reliable to estimate the effective tension in the line, especially in extreme weather conditions. In this case, dynamic models are necessary to compute the loads in the lines, and thus the restoring effect of the station-keeping system. This feature is available in some commercial modelling software such as OrcaFlex[3] [10] or ANSYS AQWA [11]. Although wave-to-wire tools may be coupled to dynamic mooring models this

---

[3]Dynamic models represent the mooring lines by a finite-element description. The equation of motion is solved at each node in order to compute the tension in the line. Consequently, the elasticity and stiffness of the line, the hydrodynamic added-mass and drag effects, and the seabed interactions, among others, can be modelled accurately. The numerical methods implemented in such codes allow making numerical predictions under extreme loads and fatigue analysis of mooring lines possible, however, the computational effort required is in general considerable. The most widespread commercial code available in the market is Orcaflex a user-friendly numerical tool that allows the user to study the most common problems in offshore industry.

is not a common approach. Wave-to-wire tools are designed to model the operational regime of WECs (i.e. during power production) where a linear (or partially non-linear) wave/device hydrodynamic interaction is fairly valid. In general, the nonlinear behaviour of WECs under extreme wave conditions is not properly represented in wave-to-wire tools. Therefore, combining wave-to-wire and dynamic mooring models does not allow the full capabilities of the dynamic model to be exploited. Moreover, usually most WECs need to enter a survival mode (with no energy production) in extreme wave conditions in order to avoid structural damage which, to some extent, decreases the usefulness of wave-to-wire models since their main feature is to assess the energy conversion efficiency.

10.2.5   Radiation Force

In addition to the usual instantaneous forces proportional to the acceleration, velocity and displacement of the body, the most commonly-used formulations of time-domain models of floating structures incorporate convolution integral terms, known as 'memory' functions. These take account of effects which persist in the free surface after motion has occurred. This 'memory' effect means that the loads on the wet body surface in a particular time instant are partially caused by the change in the pressure field induced by previous motions of the body itself. Assuming that the system is causal, this is, $h(t) = 0$ for $t < 0$, and time invariant[4] these convolution integrals take the form

$$F_r(t) = \int_0^\tau h(t - \tau)\dot{z}(\tau)d\tau, \tag{10.9}$$

where $h(t - \tau)$ represents the impulse-response functions (IRFs) or kernels of the convolutions and $\dot{z}(\tau)$ the body velocity towards any DoF. In the case of 6 rigid DoFs, $h(t - \tau)$ is a 6×6 symmetric matrix where the off-diagonal entries represent the cross-coupling radiation interaction between the different oscillatory modes.

Apart from a few cases which may be solved analytically, the IRFs are derived computationally. The most common method does not involve the direct computation of the IRFs, but derives the IRFs from the frequency-dependent hydrodynamic data obtained with standard 3D radiation/diffraction codes (such as ANSYS Aqwa [11], WAMIT [12], Moses [13] or the open source Nemoh code [14]) generally used to model WECs.

---

[4]A time-invariant system is a system whose output does not depend explicitly on time. This mathematical property may be expressed by the statement: If the input signal $x(t)$ produces an output $y(t)$ then any time shifted input, $x(t+\tau)$, results in a time-shifted output $y(t+\tau)$.

The output of these numerical tools includes the frequency dependent added mass, $A(\omega)$, and damping, $B(\omega)$, coefficients along with the added mass coefficient in the limit as the frequency tend to infinity, A_∞ (**see section: hydrodynamics**). The IRFs are normally obtained by applying the inverse discrete *Fourier* transform to the radiation transfer function, $H(\omega)$, given by

$$H(\omega) = [A_\infty - A(\omega)] + B(\omega). \tag{10.10}$$

Usually the direct computation of the convolution integrals is quite time consuming. Therefore, alternative approaches have been proposed to replace the convolution integrals in the system of motion equations, such as implementing a transfer function of the radiation convolution [15], or state-space formulations [16–18].

The state-space formulation, which originated and is generally applied in control engineering, has proved to be a very convenient technique to treat these sorts of hydrodynamic problems. Basically, this approach consists of representing the convolution integral by (ideally) a small number of first order linear differential equations with constant coefficients. For causal and time invariant systems the state-space representation is expressed by

$$\dot{X}(t) = AX(t) + B\dot{z}(t)$$

$$y(t) = \int_0^\tau h(t - \tau)\dot{z}(\tau)d\tau = CX(t), \tag{10.11}$$

where the constant coefficient array A and vectors B and C define the state-space realization and x represents the state vector, which summarizes the past information of the system at any time instant.

Different methodologies have been proposed to derive the constant coefficients of the differential equations (i.e. the array A and the vectors B and C): (i) directly from the transfer function obtained with standard hydrodynamic 3D radiation-diffraction codes or (ii) explicitly from the IRF (i.e. the Fourier transform of the transfer function). Since typically the time domain modeling of WECs involves the use of 3D radiation-diffraction codes, which give the transfer function as an output, the first alternative is more convenient and is the approach generally used as it avoids additional errors being introduced by the application of the Fourier transform to obtain the IRF.

Next, a parametric model that approximates the transfer function by a complex rational function, computed for a discrete set of frequencies, is run. The most common methodology is based on the so-called frequency response curve fitting, which seems to provide the simplest implementation method (iterative linear least squares [19, 20]). The method provides superior models, mainly if the hydrodynamic code gives the added mass at infinite frequency, because it forces the structure of the model to satisfy all the properties of the convolution terms.

The least squares approach consists of identifying the appropriate order of the numerator and denominator polynomials (rational function) and then finding the parameters of the polynomials (numerator and denominator). The parameter estimation is a non-linear least squares problem which can be linearized and solved iteratively. This operation can be performed using the *MATLAB* function *invfreqs* (signal processing toolbox) which solves the linear problem and gives as output, for a prescribed transfer function, the parameters vector [21]. To convert the transfer function filter parameters to a state-space form the signal processing toolbox of MATLAB includes the function *tf2ss,* which returns the A, B and C matrices of a state space representation for a single-input transfer function.

10.2.6 PTO Force

The simplest way to represent the PTO force involves considering a linear force that counteracts the WEC motion. This force is composed of one term proportional to the WEC velocity and another proportional to the WEC displacement, i.e,

$$F_{pto}(t) = -D\dot{z}(t) - kz(t).$$ (10.12)

The first term of Eq. 10.12 is the resistive-force component where D is the so-called damping coefficient. This term refers to a resistive or dissipative effect and is therefore related to the WEC capacity to extract wave energy. Furthermore, the second term of Eq. 10.12, represents a reactive-force proportional to the displacement, where k is the so-called spring coefficient. This term embodies a reactive effect related to the energy that flows between the PTO and the moving part of the WEC. The reactive power is related to the difference between the maximum values of kinetic and potential energy. Ultimately, the reactive-force component does not contribute to the time-averaged absorbed power since the time-averaged reactive power is zero.

To maximize the overall energy extraction (rather than the instantaneous power) it is necessary to continually adjust the characteristics of the control system in order to keep the converter operating at peak efficiency.

Fundamentally there are two main strategies to control WECs: passive control and active control. Passive control is the simplest control strategy as it consists of only applying to the floater an action proportional to its velocity (resistive force) by adjusting the damping coefficient and setting the reactive-force component of the PTO to zero. Conversely, active control requires tuning both PTO parameters, D and K, which, as mentioned above, implies bidirectional reactive power flowing between the PTO and the absorber.

Control of WECs is an intricate matter mostly due to the randomness of ocean waves and the complexity of the hydrodynamic interaction phenomenon between WECs and the ocean waves. Furthermore, an additional difficulty arises from the sensitivity of optimum control on future knowledge of the sea state (especially in the case of resonant point absorbers) [22]. However, control is crucial to enhance the system performance, particularly in the case of point absorbers where appropriate control strategies, normally highly non-linear, allow the otherwise narrow bandwith of the absorber to be broadened. In this framework the PTO machinery must have the capacity to cope with reactive forces and reactive power. Controlling the PTO reactive-force, so that the global reactance is cancelled [22], is the basis of these so called phase control methods. In this way the natural device response, including its resonant characteristics, are adjusted such that the velocity is in phase with the excitation force on the WEC, which is a necessary condition for maximum energy capture [22].

Several strategies have been suggested in the last three decades, but latching and declutching are the two most commonly used strategies categorized as phase control techniques. Latching control, originally proposed by Budal and Falnes [23], consists of blocking and dropping the captor at appropriate time instants to force the excitation force to be in phase with the buoy velocity, as described above. Extensive research has been developed in this topic, including amongst other researchers Babarit et al. [24]; Falnes and Lillebekken [25]; Korde [26] and Wright et al. [27]. Conversely, declutching control consists of manipulating the absorber motion by shifting between applying full load force or no force, allowing the absorber to move freely for periods of time. Declutching was introduced by Salter et al. [28] and latter extensively investigated by Babarit et al. [29].

The convergence into one, or possibly two or three different WECs, is still an open issue in the wave energy field. Currently there is a wide range of proposed concepts that differ on the working principle, the applied materials, the adequacy of deployment sites, and above all the type of PTO equipment and the control characteristics. Therefore, although the hydrodynamic wave/WEC interaction might be modelled using (to some extent) similar numerical approaches (independently from the technology itself), the development of generic wave-to-wire modelling tools is hampered by the wide variety of proposed PTO equipment and dissimilar control strategies, which require different modelling approaches.

Despite the number of existing PTO alternatives there are some fundamental considerations that may be made about the correlation between the type of PTO and the WEC class. In this regard it can be said that typically the PTO of OWCs consists of a turbo-generator group with an air turbine, whether Wells[5] or self-rectifying impulse turbine.[6] In the case of WECs within the class of wave-induced relative motion there are two main fundamental differences based in the amplitude of the oscillatory motion. In general the working principle of WECs with large captors and

---

[5]The Wells turbine is a low-pressure air turbine that rotates continuously in one direction in spite of the direction of the air flow. In this type of air turbine the flow across the turbine varies linearly with the pressure drop.

[6]A self-rectifying impulse turbine rotates in the same direction no matter what the direction of the airflow is, which makes this class of turbine appropriate for bidirectional airflows such as in OWC wave energy converters. In this type of air turbine the pressure-flow curve is approximately quadratic.

so high dynamic excitation loads is based on motions of very small amplitude, which typify the use of **hydraulic systems**. On the other hand, WECs with small captors (i.e. point absorbers), and so lower excitation loads, require high displacements (within certain limits) to maximize the power capture. Those concepts are, by and large, heaving resonant WECs. In this case, the most frequently used PTO equipment is direct-drive **linear generators**, where the permanent magnet and the reluctance machines are the most noteworthy systems [30].

Recently, disruptive PTO systems based on dielectric elastomer generators (DEGs) [31] have been proposed, aiming to achieve high energy conversion efficiencies, to reduce capital and operating costs, corrosion sensitivity, noise and vibration and to simplify installation and maintenance processes. However, these systems are still in a very preliminary development stage. Therefore, as the aforementioned more conventional PTO alternatives still cover most of the technologies under development; a more detailed description of those systems is presented in this section:

- **Hydraulic systems**.
  Hydraulics systems are difficult to typify because they can take many different forms. However, usually hydraulic circuits include a given number of pairs of cylinders, high-pressure and low pressure gas accumulators and a hydraulic motor. Depending on the WEC working principle the displacement of the pistons inside the cylinders is caused by the relative motion between two (or more) bodies or the relative motion between the floater and a fixed reference (e.g. sea bed). A rectifying valve assures that the liquid always enters the high-pressure accumulator and leaves the low-pressure accumulator and never otherwise, whether the relative displacement between bodies is downwards or upwards [32]. The resulting pressure difference between the accumulators, Δp_c, drives the hydraulic motor, so that the flow rate in it, Q_m, is obtained from

$$Q_m(t) = (N_c A_c)^2 G_m \Delta p_c(t), \qquad (10.13)$$

where N_c is the number of pairs of cylinders, A_c the total effective cross sectional area of a pair of cylinders and G_m a constant. The pressure difference between the accumulators, Δp_c, is given by

$$\Delta p_c(t) = \phi_h v_h(t)^{-\gamma} - \phi_l \left[\frac{V_0 - m_h v_h(t)}{m_l} \right]^{-\gamma}, \qquad (10.14)$$

where the sub-indices l and h refer to the low and high-pressure accumulators, respectively; ϕ is a constant for fixed entropy (an isentropic process is usually assumed in the modeling process), v is the specific volume of gas, γ the specific-heat ratio for the gas, m is the mass of gas, which is assumed to be unchanged during the process, and V_0 is the total volume of gas inside the accumulators, which also remains constant during the process, so that $V_0 = m_h v_h(t) = m_l v_l(t) = C^{te}$.

The total flow rate in the hydraulic circuit is given by the variation of the volume of gas inside the high-pressure accumulator, which is given by

$$Q(t) - Q_m(t) = -m_h \frac{dv_h(t)}{dt},$$ (10.15)

where Q is the volume flow rate of liquid displaced by the pistons. The useful power at a given instant, P_u, is, in any case, given by

$$P_u(t) = Q_m(t)\Delta p_c(t).$$ (10.16)

- **Air Turbines**.
  Air turbines are the natural choice for the PTO mechanism of oscillating water columns (OWCs). In essence, OWC wave energy converters consist of hollowed structures that enclose an air chamber where an internal water free surface, connected to the external wave field by a submerged aperture, oscillates. The oscillatory motion of the internal free surface, in bottom fixed structures, or the relative vertical displacement between the internal free surface and the structure, in floating concepts, causes a pressure fluctuation in the air chamber. As a result, there is an air flow moving back and forth through a turbine coupled to an electric generator.

  The Wells turbine is the most commonly used option in OWCs, whose main characteristic is the ability to constantly spin in one direction regardless of air flow direction [33]. Nevertheless, there are other alternatives such as Wells turbines with variable-pitch angle blades [34] and axial [35] or radial [36] impulse turbines. A detailed review of air turbines used in OWCs is described by Falcão and Henrriques in Ref. [37].

  To numerically model OWCs the internal surface is usually assumed to be a rigid weightless piston since the OWC's width is typically much smaller than the wavelengths of interest [38].

  The motion of the water free-surface inside the chamber, caused by the incoming waves, produces an oscillating air pressure, $p(t) + p_a$ (p_a is atmospheric pressure), and consequently displaces a mass flow rate of air through the turbine, $\dot{m}$. This is calculated from

$$\dot{m} = \frac{d(\rho V p)}{dt},$$ (10.17)

where ρ is the air density and V the chamber air volume. Often, when modeling OWCs it is also assumed that the relative variations in ρ and V are small, which is consistent with linear wave theory. In addition, ρ is commonly related to the pressure, $p + p_a$, through the linearized isentropic relation, the adequacy of which is discussed by Falcão and Justino [39]. Taking into account the previous assumptions the mass flow rate of air in Eq. 10.17 might be rewritten as

$$\dot{m} = \rho_0 q - \frac{V_0}{c_a^2} \frac{dp}{dt}, \tag{10.18}$$

where q is the volume-flow rate of air, ρ_0 and c_a are the air density and speed of sound in atmospheric conditions respectively, and V_0 is the air chamber volume in undisturbed conditions.

The mass flow rate, $\dot{m}$, can be related to the differential pressure in the pneumatic chamber, p, by means of the turbine characteristic curves. Thus applying dimensional analysis to incompressible flow turbomachinery, yields [39, 40]

$$\Phi = f_Q(\Psi), \tag{10.19}$$

$$\Pi = f_p(\Psi), \tag{10.20}$$

where Ψ is the pressure coefficient, Φ the flow coefficient and Π the power coefficient, given respectively by

$$\Psi = \frac{p}{\rho_0 N^2 D_t^2}, \tag{10.21}$$

$$\Phi = \frac{\dot{m}}{\rho_0 N D_t^3}, \tag{10.22}$$

$$\Pi = \frac{P_t}{\rho_0 N^3 D_t^5}, \tag{10.23}$$

in which ρ_0 is the air density, $N = \dot{\omega}$ the rotational speed (radians per unit time), D_t the turbine rotor diameter and P_t the turbine power output (normally the mechanical losses are ignored).

In the case of a Wells turbine, with or without guide vanes, the dimensionless relation between the flow coefficient and the pressure coefficient, Eq. 10.19, is approximately linear. Therefore Eq. 10.19 may be rewritten in the form $\Phi = K_t \Psi$, where K_t is a constant of proportionality that depends only on turbine geometry. Eventually, the relation between the mass flow rate and the pressure fluctuation can be written as

$$\dot{m} = \frac{K_t D_t}{N} p, \tag{10.24}$$

which is linear for a given turbine and constant rotational speed. The instantaneous (pneumatic) power available to the turbine is then obtained from

$$P_{\text{available}} = \frac{\dot{m}}{\rho_0} p, \tag{10.25}$$

and finally the instantaneous turbine efficiency is given by

$$\eta = \frac{P_t}{P_{\text{available}}} = \frac{\Pi}{\Phi \Psi}. \tag{10.26}$$

- **Direct drive linear generators**.
  The most typical applications of direct drive systems make use of rotating motions to convert mechanical energy into electrical energy. Generators in conventional power stations (e.g. coal, fuel oils, nuclear, natural gas), hydro power stations or direct-drive wind turbines all use rotating generators. However, in some particular cases linear generators are also used in applications with high power levels. This is the case of some hi-tech transportation systems, such as magnetic levitation (maglev) trains, and PTO systems for wave energy conversion.

The inherent complexity of extracting energy from waves, and ultimately the main difficulty with using linear generators for wave energy conversion, is related to the intricacy of handling high forces (depending on the size of the wave energy converter) and low speeds. In this context the viability of linear generators is restricted to heaving point absorbers which are characterized by higher velocities (higher that 1 m/s [41]) and lower excitation loads than the majority of the other categories of WEC. Nevertheless, the relevance of this PTO mechanism is highlighted by the large number of projects that have been focused on developing different heaving point absorber concepts equipped with linear generators (e.g. AWS, OPT, Seabased, Wedge Global, etc).

In the context of wave energy conversion there are different types of conventional linear generator that may be used. Namely

- Induction machines
- Synchronous machines with electrical excitation
- Switched reluctance machines
- Longitudinal flux permanent magnet generator.

Among these types of linear generators longitudinal flux permanent magnet generators (LFPM) have been the most common choice [41–43] for wave energy conversion. Normally, LFPM machines are also called permanent-magnet synchronous generators, as the armature winding flux and the permanent magnet flux move synchronously in the air gap. These machines have been extensively investigated for wave energy applications by Polinder and Danielsson [43, 44] amongst other researchers.

Figure 10.2 shows the cross-section of the magnetic circuit of a LFPM generator. The magnetic flux (indicated in Fig. 10.2 with dashed lines and its direction with arrows) from one magnet crosses the air gap and is conducted by the stator teeth through the stator coils. Then the flux is divided into two paths in the stator yoke and returns all the way through the stator teeth, crossing the air gap and through the adjacent magnets. The permanent magnets on the translator are mounted with alternating polarity, which creates a magnetic flux with alternating direction.

Fig. 10.2 Cross-section of a LFPM generator where the magnetic flux path is illustrated with *dashed lines* [45]

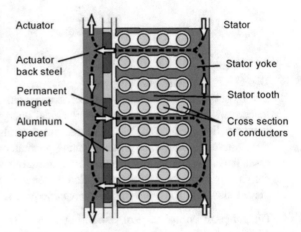

The relative motion between the stator and translator induces an electromotive force *emf* in the armature windings which drives a current whenever the armature winding is coupled to a load. In single body heaving point absorbers the translator is normally connected to the floater and the stator fixed to the sea bed, such as for the *Seabased* concept [46]. In the case of two body heaving concepts, the most common configurations have the stator attached to a submerged body and the translator connected to the floater. In turn, the current produced creates a magnetic flux that interacts with the flux of the permanent magnet leading to a force on the translator. In this way the floater mechanical energy is converted into electric energy consumed in the load.

From Faraday's law of induction the electromotive force emf, E, i.e. the voltage induced by the permanent magnet flux, may be written as

$$E = \omega \phi N, \tag{10.27}$$

where ω is the angular frequency, ϕ is the permanent magnet induced flux per pole and N is the total number of coil turns. The angular frequency is given by

$$\omega = 2\pi \frac{u_r}{w}, \tag{10.28}$$

in which u_r is the relative vertical speed between stator and translator and w the distance between the poles (i.e. the pole pitch). Simultaneously, there is also a resistive voltage drop in the slots, the end windings and cable connections when the generator is loaded. This resistive voltage drop per unit of length of the conductor is given by

$$E = I\rho_{cu},\tag{10.29}$$

where ρ_{cu} is the resistivity of the conductor material (mostly copper) and I is the current density in the conductor. As a result the induced phase currents produce a magnetic field, divided into two components: one component is coupled to the entire magnetic circuit, i.e. the main flux, and the other component is leakage flux. The corresponding inductances are then defined accordingly as the main inductance, L_m, and the leakage inductance, L_l. In a symmetric system the synchronous inductance, L_s, expressed in terms of the main inductance and the leakage inductance, is given by

$$L_s = \frac{3}{2}L_m + L_l,\tag{10.30}$$

where the first term is the armature flux linkage with the phase winding, which will be described below, and the second term is leakage inductance of that phase.

In a simplistic way the main electrical characteristics of a LFPM generator may be described using a lumped circuit as illustrated in Fig. 10.3 for a single phase of the generator. A single phase might be then modelled by an electromotive force, E, (voltage induced by the permanent magnet flux), a resistance inside the generator, R_g, a inductive voltage modelled by the synchronous inductance, L_s, and a load resistance R_l (the load might be either purely resistive or may also have a reactive component).

Fig. 10.3 Lumped circuit diagram of one phase of a synchronous generator

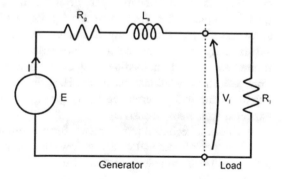

Generator   Load

From the lumped circuit we can determine the load voltage given by

$$V_l = \frac{ER_l}{R_l + R_g + i\omega L_s},\tag{10.31}$$

the phase current by

$$I = \frac{E}{R_l + R_g + i\omega L_s},$$ (10.32)

and finally the power in the load is obtained from

$$P = \frac{E^2 R_l}{\left(R_l + R_g\right)^2 + \left(\omega L_s\right)^2}.$$ (10.33)

Regardless of the type of electrical machine there are fundamentally two main electromagnetic forces: the normal force, attracting the two iron surfaces, and the thrust force, acting along the translator, in the longitudinal direction in linear machines or tangential to the rotor surface in the case of rotating generators. The corresponding sheer, τ, and normal, σ, stresses are given respectively by

$$\tau = \frac{BA_e}{2}$$ (10.34)

and

$$\sigma = \frac{B^2}{2\mu_0},$$ (10.35)

where B is the air gap magnetic flux density (the SI unit of magnetic flux density is the Tesla, denoted by T), A_e is the electrical loading, measured in amperes per metre (A/m), and μ_0 the magnetic permeability of free space, also known as the magnetic constant, measured in henries per meter (H·m^{-1}), or newton per ampere squared (NA^{-2}). Typically the shear force density, Eq. 10.34, is limited in linear machines, since the air gap flux density is limited by saturation and cannot be increased substantially in conventional machines. Moreover, the electrical loading is also limited because current loading produces heat, and heat dissipation is by and large a drawback in conventional machines. Heat dissipation can be increased to a certain extent by improving thermal design (e.g. water cooling system), but it would not be expected to increase massively.

Besides the technical requirements for operating in irregular sea conditions with very high peak forces and relatively low speeds, the design of LFPM generators has a few additional complexities related to

(i) The design of the bearing system, which is quite intricate due to the high attractive force between translator and stator.
(ii) The mechanical construction with small air gaps. The stator construction of LFPM generators is simple and robust, however typically the air gap between the stator and the rotor has to be reasonably large, which reduces the air gap flux density and so the conversion efficiency. Essentially, the size of the gap

is imposed by manufacturing tolerances, the limited stiffness of the complete construction, large attractive forces between stator and translator, thermal expansion, etc.

(iii) The power electronics converter to connect the WEC voltage (which has varying frequency and amplitude caused by the irregular motion and continuously varying speed) to the electric grid (which has fixed frequency and amplitude).

(iv) The geometry of LFPM, however, limits the stator teeth width and cross-section area of the conductors for a given pole pitch. Increasing the tooth width to increase the magnetic flux in the stator or increasing the conductor cross-section demands a larger pole pitch and the angular frequency of the flux is thus reduced. This sets a limit for the induced emf per pole and consequently the power per air pat area.

10.2.7 End Stops Mechanism

End stops are mechanisms to restrict the stroke of the WEC moving bodies in order to restrain the displacement within certain excursion limits for operational purposes, depending on the WEC working principle. End stops mechanisms are particularly important in concepts operating at high velocities (e.g. heaving point absorbers). Virtual end stops may be incorporated in wave-to-wire models either as an independent additional force, representing a physical end stop, or included in the controller in order to avoid the bodies reaching the physical end stop, or to reduce the impact when limits are reached. Control methods for handling this kind of state saturation problem consist of adding spring and/or damper (to dissipate excessive power) terms to the calculation of the machinery force set-point. For instance, this additional force may be obtained from

$$F_{es}(t) = R_m \dot{\eta} - sign(\dot{\eta}) K_{es}(|\eta| - \eta_{\lim}) H(|\eta| - \eta_{\lim}) - D_{es}\dot{\eta} u(|\eta| - \eta_{\lim}), \quad (10.36)$$

where H is the Heaviside step function and K_{es} and D_{es} are the spring and damping constants for the end stop mechanism. The constant $\eta_{\lim}$ represents the excursion for which the mechanism starts acting [47].

10.3 Benchmark Analysis

This section presents a benchmark on existing wave-to-wire models and other modeling tools, such as CFD codes, based on the Reynolds-Averaged Navier-Stokes equation (RANSE). At present CFD codes are not the most suitable tools to model the entire chain of energy conversion (at least in a

Table 10.1 Benchmark on existing WEC modeling tools

Developer	Code name	Fluid model	Hydro model	Classes of WECs	PTO model	Multi-body	Accuracy	CPU time	Sea states		
									Small to moderate	Moderate to extreme	Severe
DNV—GL[1]	Wavedyn	Perfect fluid	Linear PFT	Moving Bodies	Non-linear	√	+	+++	√	X	X
Innosea[2]	Inwave	Perfect fluid	Partially nonlinear PFT	OWC Moving Bodies	Non-linear	√	++	++	√	√	X
ECN[3]	LAMSWEC	Perfect fluid	Partially nonlinear PFT	Moving Bodies	Non-linear	X	++	++	√	√	X
ECN	ACHIL3D	Perfect fluid	Linear PFT	Moving Bodies	Non-linear	√	+	+++	√	X	X
Sandia/NREL[4]	WEC—Sim	Perfect fluid	Linear PFT	Moving Bodies	Non-linear	√	+	+++	√	X	X
Principia[5]	Diodore	Perfect fluid	Linear PFT	Moving Bodies	Non-linear	X	+	+++	√	X	X
WavEC[6]	WavEC2wire	Perfect fluid	Linear PFT	OWC Moving Bodies	Non-linear	√	+	+++	√	X	X
Marin[7]	Refresco	Viscous fluid	RANSE	Moving Bodies	N/A	√	+++	+	√	√	√
ECN	Icare	Viscous fluid	RANSE	Moving Bodies	N/A	√	+++	+	√	√	√

[1]www.gl-garradhassan.com
[2]www.innosea.fr
[3]www.ec-nantes.fr
[4]www.energy.sandia.gov
[5]www.principia.fr
[6]www.wavec.org
[7]www.marin.nl

straightforward way) and evaluate different control strategies to enhance the device performance. Nevertheless CFD codes might be extremely useful to study flow details of the wave-structure interaction (e.g. detection of flow separations, extreme loading and wave breaking).

The main differences between the codes listed in Table 10.1 reside in the theory they are based on. For instance, modelling tools based on linear potential flow theory (PFT) are not very time demanding (especially when compared with CFD codes), although they allow the representation of a non-linear configuration of the PTO mechanism, which is the most realistic scenario for the majority of wave power devices. However, these tools have a rather limited range of applicability and fairly low accuracy, largely due to the linear theory assumptions of small waves and small body motions.

Consequently, these limitations make the modelling tools based on linear potential flow theory inadequate to assess WEC survival under extreme wave loading or even throughout operational conditions when the motion of the captor is not of small amplitude. In order to overcome these limitations various models include some nonlinearities in the hydrodynamic wave-structure interaction. The most common approach consists of computing the buoyancy and Froude-Krylov excitation forces from the instantaneous position of a WEC device instead of from its mean wet surface, as considered in the traditional linear hydrodynamic approach. The major advantage of these partially nonlinear codes is widening the range of applicability from intermediate to severe sea-states.

10.4   Radiation/Diffraction Codes

Usually wave-to-wire models rely on the output from 3D radiation/diffraction codes (such as ANSYS Aqwa [11], WAMIT [12], Moses [14] or the open source Nemoh code [13]), which are based on linear (and some of them second-order) potential theory for the analysis of submerged or floating bodies in the presence of ocean waves. These sort of numerical tools use the boundary integral equation method (BIEM), also known as the panel method, to compute the velocity potential and fluid pressure on the body mean submerged surface (wetted surface in undisturbed conditions). Separate solutions for the diffraction problem, giving the effect of the incident waves on the body, and the radiation problems for each of the pre-scribed modes of motion of the bodies are obtained and then used to compute the hydrodynamic coefficients, where the most relevant are:

Added-Mass Coefficient:
The added mass is the inertia added to a (partially or completely) submerged body due to the acceleration of the mass of the surrounding fluid as the body moves through it. The added-mass coefficient may be decomposed into two terms: a frequency dependent parameter which varies in accordance to the frequency of the sinusoidal oscillation of the body and a constant term, known as the infinite added

mass, which corresponds to the inertia added to the body when its oscillatory motion does not radiate (generate) waves. This is the case when the body oscillates with "infinite" frequency or when it is submerged very deep in the water.

Damping Coefficient:

In fluid dynamics the motion of an oscillatory body is damped by the resistive effect associated with the waves generated by its motion. According to linear theory, the damping force may be mathematically modelled as a force proportional to the body velocity but opposite in direction, where the proportionality coefficient is called damping coefficient.

Excitation force coefficient:

According to linear theory the excitation coefficient is obtained by integrating the dynamic pressure exerted on the body's mean wetted surface (undisturbed body position) due to the action incident waves of unit amplitude, assuming that the body is stationary. The excitation coefficient results from adding to the integration of the pressure over the mean wetted body surface, caused by the incident wave in the absence of the body (i.e. the pressure field undisturbed by the body presence), a correction to the pressure field due to the body presence. This correction is obtained by integrating the pressure over the mean wetted body surface caused by a scattered wave owing to the presence of the body. The first term is known as the Froud-Krylov excitation and the second the scattered term.

10.5   Conclusion

Wave-to-wire models are extremely useful numerical tools for the study of the dynamic response of WECs in waves since they allow modelling of the entire chain of energy conversion from the wave-device hydrodynamic interaction to the electricity feed into the electrical grid, with a considerable high level of accuracy and relatively low CPU time. Wave-to-wire models allow the estimation of, among other parameters, the motions/velocities/accelerations of the WEC captor, structural and mooring loads, and the instantaneous power produced in irregular sea states. Therefore, these types of numerical tools are appropriate and widely used to evaluate the effectiveness of and to optimize control strategies.

Despite the usefulness of wave-to-wire models it is, however, important to bear in mind that they have some limitations that mostly arise from the linear wave theory assumptions which are usually considered in modelling the hydrodynamic interactions between ocean waves and WECs (e.g. linear waves, small response amplitudes). Although these assumptions are fairly acceptable to model the operational regime of WECs, which comprises small to moderate sea states, they are not appropriate to model the dynamic response of WECs under extreme conditions. Nevertheless, some sort of non-linear hydrodynamic modelling approaches might be included in wave to wire models (which extends the applicability of the model), such as the evaluation of the hydrostatic force at the instantaneous body position instead of at its undisturbed position and/or the non-linear description of the Froud-Krylov term in the excitation force [48]. Ultimately, it is possible to trade off accuracy and CPU time by choosing the partial non-linear hydrodynamic approach for better accuracy, or the linear approach for faster computation.

Wave-to-wire models might be also used for modelling wave energy farms instead of single isolated devices. For this purpose the model must consider additional forces on each device resulting from the waves radiated from the other devices in the wave farm. Obviously this hydrodynamic coupling effect significantly increases the CPU time. Some simplification may be considered for faster computation however, such as neglecting the effect of remote WECs, the radiation force from which tends to be irrelevant when compared with that caused by neighbouring WECs. Moreover, the farm size and the hydrodynamic coupling between the WECs manifests an additional difficulty since it makes the application of BEM codes to generate the inputs required by wave-to-wire models (matrices of hydrodynamic damping and added mass) more time consuming.

References

1. Ricci, P., Alves, M., Falcão, A., Sarmento, A.: Optimisation of the geometry of wave energy converters. In: Proceedings of the OTTI International Conference on Ocean Energy (2006)
2. Pizer, D.: The numerical prediction of the performance of a solo duck, pp. 129–137. Eur. Wave Energy Symp., Edinburgh (1993)
3. Arzel, T., Bjarte-Larsson, T., Falnes J.: Hydrodynamic parameters for a floating wec force-reacting against a submerged body. In: Proceedings of the 4th European Wave and Tidal Energy Conference (EWTEC), Denmark, pp. 267–274 (2000)
4. Structural Design of Wave Energy Devices (SDWED) project (international research alliance supported by the Danish Council for Strategic Research) (2014). www.sdwed.civil.aau.dk/software
5. Losada, I.J., Lara, J.L., Guanche, R., Gonzalez-Ondina, J.M.: Numerical analysis of wave overtopping of rubble mound breakwaters. Coast. Eng. 55(1), 47–62 (2008)
6. www.aquaret.com
7. Bogarino, B., Kofoed, J.P., Meinert, P.: Development of a Generic Power Simulation Tool for Overtopping Based WEC, p. 35. Department of Civil Engineering, Aalborg University. DCE Technical Reports; No, Aalborg (2007)
8. Harris, R.E., Johanning, L.,Wolfram, J.: Mooring Systems for Wave Energy Converters: A Review of Design Issues and Choices. Heriot-Watt University, Edinburgh, UK (2004)
9. Structural Design of Wave Energy Devices (SDWED) project (international research alliance supported by the Danish Council for Strategic Research), 2014. WP2—Moorings State of the art Copenhagen, 30 Aug 2010. www.sdwed.civil.aau.dk/
10. www.orcina.com
11. www.ansys.com/Products/Other+Products/ANSYS+AQWA
12. www.wamit.com
13. www.ultramarine.com
14. www.lheea.ec-nantes.fr/cgi-bin/hgweb.cgi/nemoh
15. Jefferys, E., Broome, D., Patel, M.: A transfer function method of modelling systems with frequency dependant coefficients. J. Guid. Control Dyn. 7(4), 490–494 (1984)
16. Yu, Z., Falnes, J.: State-space modelling of a vertical cylinder in heave. Appl. Ocean Res. 17(5), 265–275 (1995)

17. Schmiechen, M.: On state-space models and their application to hydrodynamic systems. NAUT Report 5002, Department of Naval Architecture, University of Tokyo, Japan (1973)
18. Kristansen, E., Egeland, O.: Frequency dependent added mass in models for controller design for wave motion ship damping. In: Proceedings of the 6th IFAC Conference on Manoeuvring and Control of Marine Craft, Girona, Spain (2003)
19. Levy, E.: Complex curve fitting. IRE Trans. Autom. Control AC-4, 37–43 (1959)
20. Sanathanan, C., Koerner, J.: Transfer function synthesis as a ratio of two complex polynomials. IEEE Trnas. Autom., Control (1963)
21. Perez, T., Fossen, T.: Time-domain vs. frequency-domain identification of parametric radiation force models for marine structures at zero speed. Modell. Ident. Control 29(1), 1–19 (2008)
22. Falnes, J.: Ocean Waves and Oscillating Systems. Book Cambridge University Press, Cambridge, UK (2002)
23. Budal, K., Falnes, J.: A resonant point absorber of ocean wave power. Nature 256, 478–479 (1975)
24. Babarit, A., Duclos, G., Clement, A.H.: Comparison of latching control strategies for a heaving wave energy device in random sea. App. Ocean Energy 26, 227–238 (2004)
25. Falnes, J., Lillebekken P.M.: Budals latchingcontrolled-buoy type wavepower plant. In: Proceedings of the 5th European Wave and Tidal Energy Conference (EWTEC), Cork, Irland (2003)
26. Korde, U.A.: Latching control of deep water wave energy devices using an active reference. Ocean Eng. 29, 1343–1355 (2002)
27. Wright, A., Beattie, W.C., Thompson, A., Mavrakos, S.A., Lemonis, G., Nielsen, K., Holmes, B., Stasinopoulos, A.: Performance considerations in a power take off unit based on a non-linear load. In: Proceedings of the 5th European Wave and Tidal Energy Conference (EWTEC), Cork, Irland (2003)
28. Salter, S.H., Taylor, J.R.M., Caldwell, N.J.: Power conversion mechanisms for wave energy. In: Proceedings Institution of Mechanical Engineers Part M–J. of Engineering for the Maritime Envoronment, vol. 216, pp. 1–27 (2002)
29. Babarit, A., Guglielmi, M., Clement, A.H.: Declutching control of a wave energy converter. Ocean Eng. 36, 1015–1024 (2009)
30. Santos, M., Lafoz, M., Blanco, M., García-Tabarés, L., García, F., Echeandía, A., Gavela, L.: Testing of a full-scale PTO based on a switched reluctance linear generator for wave energy conversion. In: Proceedings of the 4th International Conference on Ocean Energy (ICOE), Dublin, Irland (2012)
31. Moretti G., Fontana M., Vertechy R.: Model-based design and optimization of a dielectric elastomer power take-off for oscillating wave surge energy converters. Meccanica. Submitted to the Special Issue on Soft Mechatronics (status: in review)
32. Falcão, A.F.: Modelling and control of oscillating body wave energy converters with hydraulic power take-off and gas accumulator. Ocean Eng. 34, 2021–2032 (2007)
33. Gato, L.M.C., Falcao, A.F., Pereira, N.H.C, Pereira, R.J.: Design of wells turbine for OWC wave power plant. In: Procedings of the 1st International Offshore and Polar Engineering Conference, Edinburgh, UK (1991)
34. Setoguchi, T., Santhakumar, S., Takao, M., Kim, T.H., Kaneko, K.: A modified wells turbine for wave energy conversion. Renew. Energy 28, 79–91 (2003)
35. Maeda, H., Santhakumar, S., Setoguchi, T., Takao, M., Kinoue, Y., Kaneko, K.: Performance of an impulse turbine with fixed guide vanes for wave energy conversion. Renew. Energy 17, 533–547 (1999)
36. Pereiras, B., Castro, F., El Marjani, A., Rodriguez, M.A.: An improved radial impulse turbine for OWC. Renew. Energy 36(5), 1477–1484 (2011)
37. Falcão, A.F., Henriques, J.C.: Oscillating-water-column wave energy converters and air turbines: a review. Renew. Energy (2015). Online publication date: 1-Aug-2015
38. Evans, D.V.: The oscillating water column wave-energy device. J. Inst. Math. Appl. 22, 423–433 (1978)

39. Falcão, A.F., Justino, P.A.: OWC wave energy devices with air flow control. Ocean Eng. **26**, 1275–1295 (1999)
40. Dixon, S.L.: Fluid Mechanics and Thermodynamics of Turbomachinery, 4th edn. Butterworth, London (1998)
41. Polinder, H., Mueller, M.A., Scuotto, M., Sousa Prado, M.G.: Linear generator systems for wave energy conversion. In: Proceedings of the 7th European Wave and Tidal Energy Conference (EWTEC), Porto, Portugal (2007)
42. Polinder, H., Damen, M.E.C., Gardner, F.: Linear PM generator system for wave energy conversion in the AWS. IEEE Trans. Energy Convers. **19**, 583–589 (2004)
43. Polinder, H., Mecrow, B.C., Jack, A.G., Dickinson, P., Mueller, M.A.: Linear generators for direct drive wave energy conversion. IEEE Trans. Energy Convers. **20**, 260–267 (2005)
44. Danielsson, O., Eriksson, M., Leijon, M.: Study of a longitudinal flux permanent magnet linear generator for wave energy converters. Int. J. Energy Res. in press, available online, Wiley InterScience (2006)
45. Danielsson, O.: Wave energy conversion: linear synchronous permanent magnet generator. 102p. (Digital Comprehensive Summaries of Uppsala Dissertations from the Faculty of Science and Technology, 1651–6214; 232) (2006)
46. http://www.seabased.com/
47. Hals, J., Falnes, J., Moan, T.: A comparison of selected strategies for adaptive control of wave energy converters. J. Offshore Mech. Arct. Eng. **133**(3), 031101 (2011)
48. Gilloteaux, J.-C.: Mouvements de grande amplitude d'un corps flottant en fluide parfait. Application à la récupération de l'énergie des vagues. Ph.D thesis, Ecole Centrale de Nantes; Université de Nantes. (in French) (2007)

Erratum to: Handbook of Ocean Wave Energy

Arthur Pecher and Jens Peter Kofoed ⓘ

Erratum to:
A. Pecher and J.P. Kofoed (eds.), *Handbook of Ocean Wave Energy*, Ocean Engineering & Oceanography 7, DOI 10.1007/978-3-319-39889-1

The original version of the book was inadvertently published without the following corrections:

In Chap. 3, Fig. 3.13 was incorrect due to an error by the publisher and was replaced.

The numbering style has been changed from Chapter Content Separately to Chapter Content.

The spelling of the affiliation was corrected which should read as Aalborg University.

The erratum book has been updated with the changes.

The updated original online version for this book can be found at
10.1007/978-3-319-39889-1

A. Pecher (✉) · J.P. Kofoed
Wave Energy Research Group, Department of Civil Engineering,
Aalborg University, Aalborg, Denmark
e-mail: apecher@gmail.com

J.P. Kofoed
e-mail: jpk@civil.aau.dk